Irrigation Engineering

Irrigation Engineering

Asif Iqbal
Binuj A.R.
Deepak M

2026
Agri Horti Press
New Delhi-110 084

ISBN: 978-93-49690-35-6 (Hardbound)

Published By

Agri Horti Press

New Delhi

E-mail: agrihortipress@gmail.com

Digitally Printed at : Replika Press Pvt. Ltd.

PREFACE

Water is the cornerstone of agricultural productivity, economic stability, and sustainable development. In a world where water resources are increasingly under pressure due to population growth, climate change, and competing demands, the efficient and scientific management of irrigation systems has never been more vital. Irrigation engineering, as a specialized field, plays a critical role in ensuring that water is delivered in the right quantity, at the right time, and in the most effective manner.

This book, titled *Irrigation Engineering*, is designed to provide a comprehensive and accessible foundation in the principles, design, and management of irrigation systems. It is intended for students, engineers, agricultural professionals, and policy-makers who seek to understand the technical and practical aspects of irrigation in both traditional and modern contexts.

Beginning with an overview of the importance and history of irrigation, the book introduces the hydrological cycle and the fundamental concepts of soil-water-plant relationships. It explores various methods of irrigation—including surface, sprinkler, drip, and subsurface techniques—detailing their design parameters, operational considerations, and suitability for different crops and landscapes.

As readers progress, they will find in-depth discussions on canal systems, water conveyance and distribution structures, pumps and tubewells, water measurement techniques, and the design of field application systems. Practical challenges such as water losses, irrigation scheduling, drainage, and salinity management are also addressed, along with solutions rooted in engineering design and field practices.

Irrigation Engineering also incorporates emerging trends in water management, including the use of remote sensing, GIS, automation, and smart irrigation technologies. Sustainability and water-use efficiency are emphasized throughout, reflecting the evolving priorities of agriculture and environmental stewardship in the 21st century.

Our aim in this book is to blend theoretical knowledge with real-world application, supported by diagrams, examples, and case studies that illustrate key concepts and field scenarios. Whether used as a textbook or a reference manual, it is structured to build competence and confidence in designing and managing irrigation systems for diverse agricultural needs.

We trust that *Irrigation Engineering* will serve as a valuable guide and learning companion for all who engage with the essential task of managing water for productive and sustainable agriculture.

Asif Iqbal

Binuj A.R.

Deepak M

CONTENTS

1

IRRIGATION ENGINEERING: OVERVIEW

Irrigation Engineering is a science of planning and designing an efficient, low-cost, economic irrigation system tailored to fit natural conditions. The challenging projects of an irrigation engineer are to create watering systems in such as dams, canals, or drainage systems.

Irrigation is defined as the process of artificial supply of water to soil for raising crops. It is a science of planning and designing an efficient, low-cost, economic irrigation system tailored to fit natural conditions. It is the engineering of controlling and harnessing the various natural sources of water, by constructing dams and reservoirs, canals and headworks, and finally distributing the water to the agricultural fields. Irrigation engineering includes the study and design of works in connection with river control, drainage of waterlogged areas and generation of hydroelectric power.

The engineering of controlling and harnessing the various natural sources of water, by construction of dam and reservoirs, canals and headwork and finally distributing the water to the agriculture fields is called as irrigation engineering.

India is basically an agricultural country, and all its resources depend on the agriculture output. Water is evidently the most vital elements in the plants life.

Water is normally supplied to the plants by nature through rains. However, the total rainfall in a particular area may be either insufficient, or ill-timed.

In order to get maximum yield, it is essential to supply the optimum quantity of water, and to maintain correct timing of water. This is possible only through a systematic irrigation system by collecting water during the period of excess rainfall and releasing it to the crop as and when it is needed.

Thus the necessity of irrigation depends upon following points:

(a) Less Rainfall

(b) Non-uniform Rainfall

(c) Growing a number of crops during a year

(d) Growing perennial crops

(e) Commercial Crops with additional water

(f) Controlled water supply

Irrigation is the controlled application of water to croplands. Its primary objective is to create an optimal soil moisture regime for maximizing crop production and quality while at the same time minimizing the environmental degradation inherent in irrigation of agricultural lands. Irrigation is thus critical for food security in semi-arid and arid areas.

Irrigation has two primary objectives:

To supply the essential moisture for plant growth; &

To leach or dilute salts from the soil.

Irrigation water management deals with the frequency of irrigation, depth of water to be applied, and measures to increase the uniformity of applications. Irrigation management should be a set of practices designed to maximize efficiencies and minimize the labor and capital requirements of a particular irrigation system.

Role of Civil Engineers in Irrigation Engineering

Role of Civil Engineers in Irrigation Engineering Irrigation is a cross cutting sector that involves civil engineers, hydrologists, environmental experts, land surveyors, agricultural scientists and others, civil and irrigation engineers were important for successful irrigation schemes

The supply of water at farm turnout

Water storage in Dams

Water conveyance

Flood control

Irrigation may be defined as the science of artificial application of water to the land, in accordance with the crop requirements throughout the crop period for full-fledged nourishment of the crops. It is used to assist in the growing of agricultural crops, maintenance of landscapes, and revegetation of disturbed soils in dry areas and during periods of inadequate rainfall. Additionally, irrigation also has a few other uses in crop production, which include protecting plants against frost, suppressing weed growing in grain fields and helping in preventing soil consolidation.

In contrast, agriculture that relies only on direct rainfall is referred to as rain-fed or dryland farming. Irrigation systems are also used for dust suppression, disposal of sewage, and in mining. Irrigation is often studied together with drainage, which is the natural or artificial removal of surface and sub-surface water from a given area. Irrigation is also a term used in medical/dental fields to refer to flushing and washing out anything with water or another liquid.

Archaeological investigation has identified evidence of irrigation in Mesopotamia, Ancient Egypt and Ancient Persia (modern day Iran) as far back as the 6th millennium BCE, where barley was grown in areas where the natural rainfall was insufficient to support such a crop.

In the Zana Valley of the Andes Mountains in Peru, archaeologists found remains of three irrigation canals radiocarbon dated from the 4th millennium BCE, the 3rd millennium BCE and the 9th century CE. These canals are the earliest record of irrigation in the New World. Traces of a canal possibly dating from the 5th millennium BCE were found under the 4th millennium canal. Sophisticated irrigation and storage systems were developed by the Indus Valley Civilization in North India, including the reservoirs at Girnar in 3000 BCE and an early canal irrigation system from circa 2600 BCE. Large scale agriculture was practiced and an extensive network of canals was used for the purpose of irrigation.

There is evidence of the ancient Egyptian pharaoh Amenemhet III in the twelfth dynasty (about 1800 BCE) using the natural lake of the Faiyum Oasis as a reservoir to store surpluses of water for use during the dry seasons, the lake swelled annually from flooding of the Nile.

The Qanats, developed in ancient Persia in about 800 BCE, are among the oldest known irrigation methods still in use today. They are now found in Asia, the Middle East and North Africa. The system comprises a network of vertical wells and gently sloping tunnels driven into the sides of cliffs and steep hills to tap groundwater. The noria, a water wheel with clay pots around the rim powered by the flow of the stream (or by animals where the water source was still), was first brought into use at about this time, by Roman settlers in North Africa. By 150 BCE the pots were fitted with valves to allow smoother filling as they were forced into the water.

The irrigation works of ancient Sri Lanka, the earliest dating from about 300 BCE, in the reign of King Pandukabhaya and under continuous development for the next thousand years, were one of the most complex irrigation systems of the ancient world. In addition to underground canals, the Sinhalese were the first to build completely artificial reservoirs to store water. Due to their engineering superiority in this sector, they were often called 'masters of irrigation'. Most of these irrigation systems still exist undamaged up to now, in Anuradhapura and Polonnaruwa, because of the advanced and precise engineering. The system was extensively restored and further extended during the reign of King Parakrama Bahu (1153-1186 CE).

The oldest known hydraulic engineers of China were Sunshu Ao (6th century BCE) of the Spring and Autumn Period and Ximen Bao (5th century BCE) of the Warring States period, both of whom worked on large irrigation projects. In the Szechwan region belonging to the State of Qin of ancient China, the Dujiangyan Irrigation System was built in 256 BCE to irrigate an enormous area of farmland that today still supplies water.

By the 2nd century AD, during the Han Dynasty, the Chinese also used chain pumps that lifted water from lower elevation to higher elevation. These were powered by manual foot pedal, hydraulic waterwheels, or rotating mechanical wheels pulled by oxen. The water was used for public works of providing water for urban residential quarters and palace gardens, but mostly for irrigation of farmland canals and channels in the fields. In 15th century Korea, the world's first water gauge, uryanggye (Korean:???), was discovered in 1441. The inventor was Jang Yeong-sil, a Korean engineer of the Joseon Dynasty, under the active direction of the king, Sejong the Great. It was installed in irrigation tanks as part of a nationwide system to measure and collect rainfall for agricultural applications. With this instrument, planners and farmers could make better use of the information gathered in the survey.

In the Americas, extensive irrigation systems were created by numerous groups in prehistoric times. One example is seen in the recent archaeological excavations near the Santa Cruz River in Tucson, Arizona. They have located a village site dating from 4,000 years ago. The floodplain of the Santa Cruz River was extensively farmed during the Early Agricultural period, circa 1200 BC to AD 150. These people constructed irrigation canals and grew corn, beans, and other crops while gathering wild plants and hunting animals.

Present extent

In the middle of the 20th century, the advent of diesel and electric motors led for the first time to systems that could pump groundwater out of major aquifers faster than it was recharged. This can lead to permanent loss of aquifer capacity, decreased water quality, ground subsidence, and other problems. The future of food production in such areas as the North China Plain, the Punjab, and the Great Plains of the US is threatened.

At the global scale 2,788,000 km^2 (689 million acres) of agricultural land was equipped with irrigation infrastructure around the year 2000. About 68% of the area equipped for irrigation is located in Asia, 17% in America, 9% in Europe, 5% in Africa and 1% in Oceania. The largest contiguous areas of high irrigation density are found in North India and Pakistan along the rivers Ganges and Indus, in the Hai He, Huang He and Yangtze basins in China, along the Nile river in Egypt and Sudan, in the Mississippi-Missouri river basin and in parts of California. Smaller irrigation areas are spread across almost all populated parts of the world.

Types

Various types of irrigation techniques differ in how the water obtained from the source is distributed within the field. In general, the goal is to supply the entire field uniformly with water, so that each plant has the amount of water it needs, neither too much nor too little. The modern methods are efficient enough to achieve this goal.

Surface

In surface irrigation systems, water moves over and across the land by simple gravity flow in order to wet it and to infiltrate into the soil. Surface irrigation can be subdivided into furrow, border strip or basin irrigation. It is often called flood irrigation when the irrigation results in flooding or near flooding of the cultivated land. Historically, this has been the most common method of irrigating agricultural land.

Where water levels from the irrigation source permit, the levels are controlled by dikes, usually plugged by soil. This is often seen in terraced rice fields (rice paddies), where the method is used to flood or control the level of water in each distinct field. In some cases, the water is pumped, or lifted by human or animal power to the level of the land.

Localized

Localized irrigation is a system where water is distributed under low pressure through a piped network, in a pre-determined pattern, and applied as a small discharge to each plant or adjacent to it. Drip irrigation, spray or micro-sprinkler irrigation and bubbler irrigation belong to this category of irrigation methods.

Drip

Drip irrigation, also known as trickle irrigation, functions as its name suggests. Water is delivered at or near the root zone of plants, drop by drop. This method can be the most water-efficient method of irrigation, if managed properly, since evaporation and runoff are minimized. In modern agriculture, drip irrigation is often combined with plastic mulch, further reducing evaporation, and is also the means of delivery of fertilizer. The process is known as fertigation.

Deep percolation, where water moves below the root zone, can occur if a drip system is operated for too long of a duration or if the delivery rate is too high. Drip irrigation methods range from very high-tech and computerized to low-tech and labor-intensive. Lower water pressures are usually needed than for most other types of systems, with the exception of low energy center pivot systems and surface irrigation systems, and the system can be designed for uniformity throughout a field or for precise water delivery to individual plants in a landscape containing a mix of plant species.

Although it is difficult to regulate pressure on steep slopes, pressure compensating emitters are available, so the field does not have to be level. High-tech solutions involve precisely calibrated emitters located along lines of tubing that extend from a computerized set of valves.

Both pressure regulation and filtration to remove particles are important. The tubes are usually black (or buried under soil or mulch) to prevent the growth of algae and to protect the polyethylene from degradation due to ultraviolet light. But drip irrigation can also be as low-tech as a porous clay vessel sunk into the soil and occasionally filled from a hose or bucket. Subsurface drip irrigation has been used successfully on lawns, but it is more expensive than a more traditional sprinkler system.

Surface drip systems are not cost-effective (or aesthetically pleasing) for lawns and golf courses. In the past one of the main disadvantages of the subsurface drip irrigation (SDI) systems, when used for turf, was the fact of having to install the plastic lines very close to each other in the ground, therefore disrupting the turf grass area. Recent technology developments on drip installers like the drip installer at New Mexico State University Arrow Head Center, places the line underground and covers the slit leaving no soil exposed.

Sprinkler

In sprinkler or overhead irrigation, water is piped to one or more central locations within the field and distributed by overhead high-pressure sprinklers or guns. A system utilizing sprinklers, sprays, or guns mounted overhead on permanently installed risers is often referred to as a solid-set irrigation system. Higher pressure sprinklers that rotate are called rotors and are driven by a ball drive, gear drive, or impact mechanism.

Rotors can be designed to rotate in a full or partial circle. Guns are similar to rotors, except that they generally operate at very high pressures of 40 to 130 lbf/in^2 (275 to 900 kPa) and flows of 50 to 1200 US gal/min (3 to 76 L/s), usually with nozzle diameters in the range of 0.5 to 1.9 inches (10 to 50 mm). Guns are used not only for irrigation, but also for industrial applications such as dust suppression and logging.

Sprinklers can also be mounted on moving platforms connected to the water source by a hose. Automatically moving wheeled systems known as traveling sprinklers may irrigate areas such as small farms, sports fields, parks, pastures, and cemeteries unattended. Most of these utilize a length of polyethylene tubing wound on a steel drum. As the tubing is wound on the drum powered by the irrigation water or a small gas engine, the sprinkler is pulled across the field. When the sprinkler arrives back at the reel the system shuts off.

This type of system is known to most people as a "waterreel" traveling irrigation sprinkler and they are used extensively for dust suppression, irrigation, and land application of waste water. Other travelers use a flat rubber hose that is dragged along behind while the sprinkler platform is pulled by a cable. These cable-type travelers are definitely old technology and their use is limited in today's modern irrigation projects.

Center pivot

Center pivot irrigation is a form of sprinkler irrigation consisting of several segments of pipe (usually galvanized steel or aluminum) joined together and supported by trusses, mounted on wheeled towers with sprinklers positioned along its length. The system moves in a circular pattern and is fed with water from the pivot point at the center of the arc. These systems are found and used in all parts of the nation and allow irrigation of all types of terrain. Newer irrigations have drops as shown in the image that follows.

Most center pivot systems now have drops hanging from a u-shaped pipe attached at the top of the pipe with sprinkler heads that are positioned a few feet (at most) above the crop, thus limiting evaporative losses.

Drops can also be used with drag hoses or bubblers that deposit the water directly on the ground between crops. Crops are often planted in a circle to conform to the center pivot. This type of system is known as LEPA (Low Energy Precision Application). Originally, most center pivots were water powered. These were replaced by hydraulic systems (T-L Irrigation) and electric motor driven systems (Reinke, Valley, Zimmatic). Many modern sprinklers features GPS devices.

Lateral move (side roll, wheel line)

A series of pipes, each with a wheel of about 1.5 m diameter permanently affixed to its midpoint and sprinklers along its length, are coupled together at one edge of a field. Water is supplied at one end using a large hose. After sufficient water has been applied, the hose is removed and the remaining assembly rotated either by hand or with a purpose-built mechanism, so that the sprinklers move 10 m across the field.

The hose is reconnected. The process is repeated until the opposite edge of the field is reached. This system is less expensive to install than a center pivot, but much more labor intensive to operate, and it is limited in the amount of water it can carry.

Most systems utilize 4 or 5-inch (130 mm) diameter aluminum pipe. One feature of a lateral move system is that it consists of sections that can be easily disconnected. They are most often used for small or oddly shaped fields, such as those found in hilly or mountainous regions, or in regions where labor is inexpensive.

Sub-irrigation

Subirrigation also sometimes called seepage irrigation has been used for many years in field crops in areas with high water tables. It is a method of artificially raising the water table to allow the soil to be moistened from below the plants' root zone.

Often those systems are located on permanent grasslands in lowlands or river valleys and combined with drainage infrastructure. A system of pumping stations, canals, weirs and gates allows it to increase or decrease the water level in a network of ditches and thereby control the water table.

Sub-irrigation is also used in commercial greenhouse production, usually for potted plants. Water is delivered from below, absorbed upwards, and the excess collected for recycling. Typically, a solution of water and nutrients floods a container or flows through a trough for a short period of time, 10-20 minutes, and is then pumped back into a holding tank for reuse.

Sub-irrigation in greenhouses requires fairly sophisticated, expensive equipment and management. Advantages are water and nutrient conservation, and labor-saving through lowered system maintenance and automation. It is similar in principle and action to subsurface drip irrigation.

Manual using buckets or watering cans

These systems have low requirements for infrastructure and technical equipment but need high labor inputs. Irrigation using watering cans is to be found for example in peri-urban agriculture around large cities in some African countries.

Automatic, non-electric using buckets and ropes

Besides the common manual watering by bucket, an automated, natural version of this also exist. Using plain polyester ropes combined with a prepared ground mixture can be used to water plants from a vessel filled with water. The ground mixture would need to be made depending on the plant itself, yet would mostly consist of black potting soil, vermiculite and perlite. This system would (with certain crops) allow to save expenses as it does not consume any electricity and only little water (unlike sprinklers, water timers, ...). However, it may only be used with certain crops (probably mostly larger crops that do not need a humid environment; perhaps e.g. paprikas).

Using water condensed from humid air

In countries where at night, humid air sweeps the countryside, water can be obtained from the humid air by condensation onto cold surfaces. This is for example practiced in the vineyards at Lanzarote using stones to condense water or with various fog collectors based on canvas or foil sheets.

Sources of irrigation water

Sources of irrigation water can be groundwater extracted from springs or by using wells, surface water withdrawn from rivers, lakes or reservoirs or non-conventional sources like treated wastewater, desalinated water or drainage water. A special form of irrigation using surface water is spate irrigation, also called floodwater harvesting. In case of a flood (spate) water is diverted to normally dry river beds (wadis) using a network of dams, gates and channels and spread over large areas. The moisture stored in the soil will be used thereafter to grow crops.

Spate irrigation areas are in particular located in semi-arid or arid, mountainous regions. While floodwater harvesting belongs to the accepted irrigation methods, rainwater harvesting is usually not considered as a form of irrigation. Rainwater harvesting is the collection of runoff water from roofs or unused land and the concentration of this. Some of Ancient India's water systems were pulled by oxen.

WATER SCARCITY

Fifty years ago, the common perception was that water was an infinite resource. At this time, there were fewer than half the current number of people on the planet.

People were not as wealthy as today, consumed fewer calories and ate less meat, so less water was needed to produce their food. They required a third of the volume of water we presently take from rivers. Today, the competition for water resources is much more intense. This is because there are now nearly seven billion people on the planet, their consumption of water-thirsty meat and vegetables is rising, and there is increasing competition for water from industry, urbanisation and biofuel crops. To avoid a global water crisis, farmers will have to strive to increase productivity to meet growing demands for food, while industry and cities find ways to use water more efficiently.

Successful agriculture is dependent upon farmers having sufficient access to water. However, water scarcity is already a critical constraint to farming in many parts of the world. Physical water scarcity is where there is not enough water to meet all demands, including that needed for ecosystems to function effectively. Arid regions frequently suffer from physical water scarcity. It also occurs where water seems abundant but where resources are over-committed. This can happen where there is over development of hydraulic infrastructure, usually for irrigation.

Symptoms of physical water scarcity include environmental degradation and declining groundwater. Economic scarcity, meanwhile, is caused by a lack of investment in water or insufficient human capacity to satisfy the demand for water. Symptoms of economic water scarcity include a lack of infrastructure, with people often having to fetch water from rivers for domestic and agricultural uses. Some 2.8 billion people currently live in water-scarce areas.

How an in-ground irrigation system works

Most commercial and residential irrigation systems are "in ground" systems, which means that everything is buried in the ground. With the pipes, sprinklers, emitters (drippers), and irrigation valves being hidden, it makes for a cleaner, more presentable landscape without garden hoses or other items having to be moved around manually. This does, however, create some drawbacks in the maintenance of a completely buried system.

Water source and piping

The beginning of a sprinkler system is the water source. This is usually a tap into an existing (city) water line or a pump that pulls water out of a well or a pond. The water travels through pipes from the water source through the valves to the sprinklers and emitters. The pipes from the water source up to the irrigation valves are called "mainlines", and the lines from the valves to the emitters or sprinklers are called "lateral lines".

Most piping used in irrigation systems today are HDPE and MDPE or PVC or PEX plastic pressure pipes due to their ease of installation and resistance to corrosion. After the water source, the water usually travels through a check valve. This prevents water in the irrigation lines from being pulled back into and contaminating the clean water supply. Ideally a pressure control valve is also installed to regulate water pressure and help prevent excessive pressure from harming the system.

Controllers, zones, and valves

Most irrigation systems are divided into zones. A zone is a single irrigation valve and one or a group of drippers or sprinklers that are connected by pipes or tubes. Irrigation systems are divided into zones because there is usually not enough pressure and available flow to run sprinklers for an entire yard or sports field at once. Each zone has a solenoid valve on it that is controlled via wire by an irrigation controller. The irrigation controller is either a mechanical (now the "dinosaur" type) or electrical device that signals a zone to turn on at a specific time and keeps it on for a specified amount of time. "Smart Controller" is a recent term used to describe a controller that is capable of adjusting the watering time by itself in response to current environmental conditions. The smart controller determines current conditions by means of historic weather data for the local area, a soil moisture sensors (water potential or water content), rain sensor, or in more sophisticated systems satellite feed weather station, or a combination of these.

Emitters and sprinklers

When a zone comes on, the water flows through the lateral lines and ultimately ends up at the irrigation emitter (drip) or sprinkler heads. Many sprinklers have pipe thread inlets on the bottom of them which allows a fitting and the pipe to be attached to them. The sprinklers are usually installed with the top of the head flush with the ground surface.

When the water is pressurized, the head will pop up out of the ground and water the desired area until the valve closes and shuts off that zone. Once there is no more water pressure in the lateral line, the sprinkler head will retract back into the ground. Emitters are generally laid on the soil surface or buried a few inches to reduce evaporation losses.

Problems in irrigation

Irrigation can lead to a number of problems:

- Competition for surface water rights.
- Depletion of underground aquifers.

- Ground subsidence (e.g. New Orleans, Louisiana)
- Under-irrigation or irrigation giving only just enough water for the plant (e.g. in drip line irrigation) gives poor soil salinity control which leads to increased soil salinity with consequent build up of toxic salts on soil surface in areas with high evaporation. This requires either leaching to remove these salts and a method of drainage to carry the salts away. When using drip lines, the leaching is best done regularly at certain intervals (with only a slight excess of water), so that the salt is flushed back under the plant's roots.
- Over-irrigation because of poor distribution uniformity or management wastes water, chemicals, and may lead to water pollution.
- Deep drainage (from over-irrigation) may result in rising water tables which in some instances will lead to problems of irrigation salinity requiring watertable control by some form of subsurface land drainage.
- Irrigation with saline or high-sodium water may damage soil structure owing to the formation of alkaline soil

Deficit irrigation

Deficit irrigation (DI) is a watering strategy that can be applied by different types of irrigation application methods. The correct application of DI requires thorough understanding of the yield response to water (crop sensitivity to drought stress) and of the economic impact of reductions in harvest. In regions where water resources are restrictive it can be more profitable for a farmer to maximize crop water productivity instead of maximizing the harvest per unit land. The saved water can be used for other purposes or to irrigate extra units of land. DI is sometimes referred to as incomplete supplemental irrigation or regulated DI.

Deficit irrigation (DI) has been reviewed and defined as follows:

"Deficit irrigation is an optimization strategy in which irrigation is applied during drought-sensitive growth stages of a crop. Outside these periods, irrigation is limited or even unnecessary if rainfall provides a minimum supply of water. Water restriction is limited to drought-tolerant phenological stages, often the vegetative stages and the late ripening period.

Total irrigation application is therefore not proportional to irrigation requirements throughout the crop cycle. While this inevitably results in plant drought stress and consequently in production loss, DI maximizes irrigation water productivity, which is the main limiting factor (English, 1990). In other words, DI aims at stabilizing yields and at obtaining maximum crop water productivity rather than maximum yields (Zhang and Oweis, 1999)."

CROP WATER PRODUCTIVITY

Crop water productivity (WP) or water use efficiency (WUE) expressed in kg/m^3 is an efficiency term, expressing the amount of marketable product (e.g. kilograms of grain) in relation to the amount of input needed to produce that output (cubic meters of water). The water used for crop production is referred to as crop evapotranspiration. This is a combination of water lost by evaporation from the soil surface and transpiration by the plant, occurring simultaneously. Except by modeling, distinguishing between the two processes is difficult. Representative values of WUE for cereals at field level, expressed with evapotranspiration in the denominator, can vary between 0.10 and 4 kg/m3.

Experiences with deficit irrigation

For certain crops, experiments confirm that DI can increase water use efficiency without severe yield reductions. For example for winter wheat in Turkey, planned DI increased yields by 65% as compared to winter wheat under rainfed cultivation, and had double the water use efficiency as compared to rainfed and fully irrigated winter wheat. Similar positive results have been described for cotton.

Experiments in Turkey and India indicated that the irrigation water use for cotton could be reduced to up to 60 percent of the total crop water requirement with limited yield losses. In this way, high water productivity and a better nutrient-water balance was obtained.

Certain Underutilized and horticultural crops also respond favorably to DI, such as tested at experimental and farmer level for the crop quinoa. Yields could be stabilized at around 1.6 tons per hectare by supplementing irrigation water if rainwater was lacking during the plant establishment and reproductive stages. Applying irrigation water throughout the whole season (full irrigation) reduced the water productivity. Also in viticulture and fruit tree cultivation, DI is practiced.

Scientists affiliated with the Agricultural Research Service (ARS) of the USDA found that conserving water by forcing drought (or deficit irrigation) on peanut plants early in the growing season has shown to cause early maturation of the plant yet still maintain sufficient yield of the crop. Inducing drought through deficit irrigation earlier in the season caused the peanut plants to physiologically "learn" how to adapt to a stressful drought environment, making the plants better able to cope with drought that commonly occurs later in the growing season.

Deficit irrigation is beneficial for the farmers because it reduces the cost of water and prevents a loss of crop yield (for certain crops) later on in the growing season due to drought. In addition to these findings, ARS scientists suggest that deficit irrigation accompanied with conservation tillage would greatly reduce the peanut crop water requirement.

For other crops, the application of deficit irrigation will result in a lower water use efficiency and yield. This is the case when crops are sensitive to drought stress throughout the complete season, such as maize.

Apart from university research groups and farmers associations, international organizations such as FAO, ICARDA, IWMI and the CGIAR Challenge Program on Water and Food are studying DI.

Reasons for increased water productivity under deficit irrigation

If crops have certain phenological phases in which they are tolerant to water stress, DI can increase the ratio of yield over crop water consumption (evapotranspiration) by either reducing the water loss by unproductive evaporation, and/or by increasing the proportion of marketable yield to the totally produced biomass (harvest index), and/or by increasing the proportion of total biomass production to transpiration due to hardening of the crop - although this effect is very limited due to the conservative relation between biomass production and crop transpiration, - and/or due to adequate fertilizer application and/or by avoiding bad agronomic conditions during crop growth, such as water logging in the root zone, pests and diseases, etc.

Advantages

The correct application of DI for a certain crop:

- Maximizes the productivity of water, generally with adequate harvest quality;
- Allows economic planning and stable income due to a stabilization of the harvest in comparison with rainfed cultivation;
- Decreases the risk of certain diseases linked to high humidity (e.g. fungi) in comparison with full irrigation;
- Reduces nutrient loss by leaching of the root zone, which results in better groundwater quality and lower fertilizer needs as for cultivation under full irrigation;
- Improves control over the sowing date and length of the growing period independent from the onset of the rainy season and therefore improves agricultural planning.

Constraints

A number of constraints apply to deficit irrigation:

- Exact knowledge of the crop response to water stress is imperative.
- There should be sufficient flexibility in access to water during periods of high demand (drought sensitive stages of a crop).
- A minimum quantity of water should be guaranteed for the crop, below which DI has no significant beneficial effect.
- An individual farmer should consider the benefit for the total water users community (extra land can be irrigated with the saved water), when he faces a below-maximum yield;
- Because irrigation is applied more efficiently, the risk for soil salinization is higher under DI as compared to full irrigation.

MODELING

Field experimentation is necessary for correct application of DI for a particular crop in a particular region. In addition, simulation of the soil water balance and related crop growth (crop water productivity modeling) can be a valuable decision support tool. By conjunctively simulating the effects of different influencing factors (climate, soil, management, crop characteristics) on crop production, models allow to (1) better understand the mechanism behind improved water use efficiency, to (2) schedule the necessary irrigation applications during the drought sensitive crop growth stages, considering the possible variability in climate, to (3) test DI strategies of specific crops in new regions, and to (4) investigate the effects of future climate scenarios or scenarios of altered management practices on crop production.

Irrigation in viticulture

The role of irrigation in viticulture is considered both controversial and essential to wine production. In the physiology of the grapevine, water is a vital component to function of the vine with its presence or lack impacting photosynthesis, new plant shoot growth, as well as the development of grape berries. While climate and humidity play important roles, a typical grape vine needs 25-35 inches (635-890 millimeters) of water a year, occurring during the spring and summer months of the growing season, to avoid stress. A vine that does not receive the necessary amount of water will have its growth altered in a number of ways; some effects of water stress (particularly, smaller berry size and somewhat higher sugar content) are considered desirable by wine grape growers.

In many Old World wine regions, natural rainfall is considered the only source for water that will still allow the vineyard to maintain its terroir characteristics. The practice of irrigation is viewed by some critics as unduly manipulative with the potential for detrimental wine quality due to high yields that can be artificially increased with irrigation. It has been historically banned by the European Union's wine laws, though in recent years individual countries (such as Spain) have been loosening their regulations and France's wine governing body, the Institut National des Appellations d'Origine (INAO), has also been reviewing the issue.

In very dry climates that receive little rainfall, irrigation is considered essential to any viticultural prospects. Many New World wine regions such as Australia and California regularly practice irrigation in areas that couldn't otherwise support viticulture. Advances and research in these wine regions (as well as some Old World wine regions such as Israel), have shown that potential wine quality could increase in areas where irrigation is kept to a minimum and managed. The main principle behind this is controlled water stress where the vine receives sufficient water during the budding and flowering period that is then scaled back during the ripening period where the vine then responds by funneling more its limited resources into developing the grape clusters instead of excess foliage. If the vine receives too much water stress, then photosynthesis and other important process could be impacted with vine essentially shutting down. The availability of irrigation means that if drought conditions emerge, sufficient water can be provided for the plant so that the balance between water stress and development is kept to optimal levels.

The practice of irrigation has had a long history in wine production. Archaeologists describe it as one of the oldest practices in viticulture, with irrigation canals discovered near vineyard sites in Armenia and Egypt dating back more than 2600 years. Irrigation was already widely practice for other agricultural crops since around 5000 BC. It is possible that the knowledge of irrigation helped viticulture spread from these areas to other regions due to the potential for the grapevine to grow in soils too infertile to support other food crops. A somewhat hardy plant, the grapevine largest need is for sufficient sunshine and is able to flourish with minimum needs of water and nutrients. In areas where its water needs are unfulfilled, the availability of irrigation meant that viticulture could still be supported.

In the 20th century, the growing wine industries of California, Australia and Israel were greatly enhanced by advances in irrigation. With the development of more cost efficient and less labor intensive ways of watering the vines, vast tracks of very sunny but dry lands were able to be converted into growing wine regions. The ability to control the precise amount of water each vine received, allowed producers in these New World wine regions to develop styles of wines that could be fairly consistent each year regardless of normal vintage variation.

This created a stark contrast to the Old World wine regions of Europe where vintage variation, including rainfall, had a pronounced effect on the potential wine style each year. Continuing research explored the way that controlled (or supplemental) irrigation could be used to increase potential wine quality by influence how the grapevine responds to its environment and funnels resources into developing the sugars, acids and phenolic compounds that contribute to a wine's quality. This research lead to the development of ways to measure the amount water retention in the soil to where individual irrigation regimes could be plotted for each vineyard that maximized the benefits of water management.

Role of water in viticulture

The presence of water is essential for the survival of all plant life. In a grapevine, water acts as a universal solvent for many of the nutrients and minerals needed to carry out important physiological functions-which the vine receives by absorbing the nutrient-containing water from the soil. In the absence of water in the soil, the root system of the vine may have difficulties absorbing these nutrients.

Within the structure of the plant itself, water acts as a transport within the xylem bring these nutrients to all ends of the plant. During the process of photosynthesis, water molecules combine with carbon derived from carbon dioxide to form glucose which is the primary energy source of the vine as well as oxygen as a by-product.

In addition to its use in photosynthesis, a vine's water supply is also depleted by the processes of evaporation and transpiration. In evaporation, heat (aided by wind and sunlight) causes water moisture in the soil to evaporate and escape as vapor molecules. This process is inversely related to humidity with evaporation often taking place at faster rates in areas with low relative humidity.

In transpiration, this evaporation of water occurs directly in the wine as water is released from the plant through the stomata that is located underneath the leafs of a grape vine. This process helps the vine combat against the effects of heat stress which can severely damage the physiological functions of the vine (somewhat similar to how perspiration works with humans and animals).

The presence of adequate water in the vines can help keep the internal temperature of a the leaf only a few degrees above the temperature of the surrounding air. However, if water is severely lacking then that internal temperature could jump nearly 18 °F (10 °C) warmer than the surrounding air which leads the vine to develop heat stress. The dual effects of evaporation and transpiration are called evapotranspiration. A typical vineyard in a hot, dry climate can lose as much as 1,700 U.S. gallons (6,400 l; 1,400 imp gal) of water per vine through evapotranspiration during the growing season.

Factors influencing irrigation

Climates with low humidity promote faster rates evapotranspiration which reduce the grapevine's water supply. These areas may need to utilize supplemental irrigation.

There are essentially two main types irrigation-primary irrigation, which is needed for areas (such as very dry climates) that lack sufficient rainfall for viticulture to even exist, and supplemental irrigation where irrigation is used to "fill in the gaps" of natural rainfall to bring water levels to better numbers as well as to serve as a preventive measure in case of seasonal drought conditions. In both cases, both the climate and the vineyard soils of the region will play an instrumental role in irrigation's use and effectiveness.

Impact of different climate types

Viticulture is most commonly found in Mediterranean, continental and maritime climates with each unique climate providing its own challenges in providing sufficient water at critical times during the growing season. In Mediterranean climates irrigation is usually needed during the very dry periods of the summer ripening stages where drought can be a persistent threat. The level of humidity in a particular macro-climate will dictate exactly how much irrigation is needed with high levels of evapotranspiration more commonly occurring in Mediterranean climates that have low levels of humidity such as part of Chile and the Cape Province of South Africa.

In these low humidity regions, primary irrigation may be needed, but in many Mediterranean climates the irrigation is usually supplemental. The amount of precipitation that occurs during spring and summer months is also important. For example, Tuscany receives an average of 8 inches (200 mm) of rainfall during the months of April through June - the period that includes flowering and fruit set, when the water is most crucial.

While fluctuations in rainfall do occur, the amount of natural precipitation, combined with water holding capacity of soil, is typically sufficient to result in healthy harvest. In contrast, Napa Valley only gets 2.4 inches (60 mm), on average, during the same period of time, often in an erratic pattern (some years seeing more, some years seeing only trace amounts of rain), and most appellations in Central and Southern California (both along the coast and inland) receive even less than that, necessitating supplemental irrigation.

Continental climates are usually seen in areas further inland from the coastal influences of oceans and large bodies of water. The difference from the average mean temperature of its coldest and hottest months can be quite significant with moderate precipitation that usually occurs in the winter and early spring.

Depending on the water retaining ability of the soil the grapevine may receive enough water during this period to last throughout the growing season with little if any irrigation needed. For soils with poor water retention, the dry summer months may require some supplemental irrigation. Examples of continental climates that use supplemental irrigation include the Columbia Valley of Washington State and the Mendoza wine region of Argentina.

Maritime climates tend to fall between Mediterranean and continental climates with a moderate climate that is tempered by the effects of a large body of water nearby. As with Mediterranean climates, the humidity of the particular macroclimate will play a significant role in determining how much irrigation is needed. In most cases irrigation, if it is used at all, will only be supplemental in years where drought may be an issue. Many maritime regions, such as Rias Baixas in Galicia, Bordeaux and the Willamette Valley in Oregon, suffer from the diametric problem of having too much rain during the growing season.

Impact of different soil types

Soil can have a significant impact on the potential quality of wine. While geologist and viticulturist are not exactly sure what type of immutable or terroir based qualities that soil can impart on wine, there is near universal agreement that a soil's water retention and drainage abilities play a primary role. Water retention refers to the soil's ability to hold water. The term "field capacity" is used to describe the maximum amount of water that deeply moisten soil will retain after normal drainage.

Drainage is the ability of water to move freely throughout the soil. The ideal circumstance is soil that can retain sufficient amount of water for the grapevine but drains well enough to where the soil doesn't become water-logged. Soil that doesn't retain water well encourages the vine to easily sleep into water stress while soil that doesn't drain well runs of the risk of water-logged roots being attacked by microbial agents that consume all the soil nutrients and end up starving the vine.

The depth, texture and composition of soils can influence its water retaining and draining ability. Soils containing large amounts of organic material tend to have the highest water retention abilities. These types of soils include deep loams, silty soils like what is typically found on the fertile valley floors such as in the California's Napa Valley. Clay particles have the potential to remain in colloidal suspension for long periods of time when it is dissolved in water. This gives clay-based soils the potential to retain significant amount of water such as the clay soils of the Right bank Bordeaux region of Pomerol.

Many regions with these types of water retaining soils have little need for irrigation, or if they do it is usually supplemental during periods of drought. Soils with poor water retention include sand and alluvial gravel based soils such as those found in the Barolo and Barbaresco zones of Italy or in many areas of South Australia. Depending on the climate and amount of natural rainfall, areas with poor water retention may need irrigation.

Just as having too little water is detrimental to the grapevine, so too is having too much. When vines become water-logged they become a target for various microbial agents such as bacteria and fungi that compete with the vine for nutrients in the soil. Additionally excessively moist soil is poor conductor of valuable heat radiating from the ground. In general wet soils are cold soils which can be especially problematic during the flowering causing poor berry set that could lead to coulure.

It also becomes an issue during the ripening stage when vines in cool-climate regions may need additional heat radiated from the ground in order to sufficiently ripen its fruit (an example of this is the slate-based vineyards of the Mosel in Germany). Therefore, well draining soils are considered very conducive to producing quality wine. In general light-textured (such as sand and gravel) and stony soils tend to drain well.

Soils heavy soils and those with high proportions of organic matter also have the potential to drain well if they having a crumbling texture and structure. This texture relates to the friability of the soil which can come from earthworms and other organisms that have burrowed tunnels throughout the soil. Much like rocks, these tunnels give water a freer passageway through soil and contributes to its drainage.

Measuring soil moisture

Tensiometers can be used to measure soil moisture. The components of this example include (1) porous cup, (2) water-filled tube (3) sensor-head and a (4) pressure sensor.

Because of the problems associated with water-logged and wet soils, it is important for viticulturist to know how much water is currently in the soil before deciding if and how much to irrigate. There are several methods of evaluating soil moisture. The most basic is simple observation and feeling of the soil, however this has its limitations since the subsoil may be moist while the surface soil appears dry. More specific measurements can be attained by using tensiometers which evaluates surface tension of water extracted from the soil.

The presence of water in the soil can be measured by neutron moisture meters that utilize an aluminium tube with an internal neutron source that detect the subtle change between the water in the soil. Similarly, gypsum block placed throughout the vineyard contain an electrode that can be used to detect the electrical resistance that occurs as the soil dries and water is released by evaporation. Since the 1990s there has been greater research into tools utilizing time-domain reflectometry and capacitance probes. In addition to monitoring for excessive moisture, viticulturists also keep an eye for signs of water stress (discussed below) due to severe lack of water.

Irrigation systems

There are several methods of irrigation that can be used in viticulture depending on the amount of control and water management desired. Historically, surface irrigation was the most common means using the gravity of a slope to release a flood of water across the vineyard. In the early history of the Chilean wine industry, flood irrigation was widely practiced in the vineyards using melted snow from the Andes Mountains channeled down to the valleys below.

This method provided very little control and often had the adverse effect of over-watering the vine. An adaption of method was the furrow irrigation system used in Argentina where small channels ran through the vineyard providing irrigation. This provide slightly more control since the initial amount of water entering the channels could be regulated, however the amount that each vine received was sporadic.

Sprinkler irrigation involves the installation of a series of sprinkler units throughout the vineyard, often spaced as several rows about 65 feet (20 m) apart. The sprinklers can be set on a electronic timer and release predetermined amount of water for a set period of time. While this provides more control and uses less water than flood irrigation, like furrow irrigation the amount that each individual wine receives can be sporadic. The irrigation system that provides the most control over water management, though conversely the most expensive to install, is drip irrigation.

This system involved long plastic water supply lines that run down each row of vines in the vineyard with each individual grape vine having its own individual dripper. With this system, a viticulturist can control the precise amount of water that each grapevine gets down to the drop. An adaption of this system, potentially useful in areas where irrigation may be banned, is underground sub-irrigation where precise measurements of water is delivered directly to the root system.

When and how much?

With abundant water, a grapevine will produce shallow root systems and vigorous growths of new plant shoots. This can contribute to a large, leafy canopy and high yields of large grape berry clusters that may not be sufficiently or physiologically ripe. With insufficient water, many of the vine's important physiological structures, including photosynthesis that contributes to the development of sugars and phenolic compounds in the grape, can shut down.

The key to irrigation is to provide just enough water for the plant to continuing function without encouraging vigorous growth of new shoots and shallow roots. The exact amount of water will depend on a variety of factors including how much natural rainfall can be expected as well as the water retaining and drainage properties of the soil.

Water is very crucial during the early budding and flowering stages of the growing season. In areas where there is not sufficient rainfall, irrigation may be needed during this time in the spring. After fruit set, the water needs for the vine drop and irrigation is often withheld till the period of veraison when the grapes begin to change color.

This period of "water stress" encourages the vine to concentrate its limited resources into lower yields of smaller berries creating a favorable skin to juice ratio that is often desirable in quality wine production. The benefits or disadvantageous of irrigation during the ripening period itself is a matter of debate and continuing research in the wine growing community.

The only area of mostly agreement is the disadvantages of water close to harvest after a prolong dry period. Grapevines that has been subjugated to prolong water stress have a tendency to rapidly absorb large amounts of water if its provided. This will dramatically swell the berries, potentially causing to them crack or burst which will make the prone to various grape diseases. Even if the berries do not crack or burst, the rapid swelling of water will cause a reduce concentration in sugars and phenolic compounds in the grape producing wines with diluted flavors and aromas.

Water stress

One of the goals of controlled, mild water stress is to discourage the formation of excess new plant growths (a bud pictured) which will compete with the developing grape clusters for the vine's limited resources. The term water stress describes the physiological states that grapevines experience when they are deprived of water. When a grapevine goes into water stress one of its first functions is to reduce the growth of new plant shoots which compete with the grape clusters for nutrients and resources.

The lack of water also keeps the individual grape berries down to a smaller size which increase its skin to juice ratio. As the skin is filled with color phenols, tannin and aroma compounds, the increase skin to juice ratio is desirable for the potential added complexity the wine may have.

While there is disagreement over exactly how much water stress if beneficial in development grapes for quality wine production, most viticulturist agree that some water stress can be beneficial. The grapevines in many Mediterranean climates such as Tuscany in Italy and the Rhone Valley in France experience natural water stress due to the reduce rainfall that occurs during the summer growing season.

At the far extreme is severe water stress which can have detrimental effects on both the vine and on potential wine quality. To conserve water, a vine will try to conserve water by limiting its lose through transpiration. The plant hormone abscisic acid triggers the stomata on the underside of the plant leaf to stay close in order to reduce the amount of water that is evaporated. While conserving water this also has the consequences of limiting the intake of carbon dioxide needed to sustain photosynthesis.

If the vine is continually stressed it will keeps it stomata closed for longer and longer periods of time which can eventually cause photosynthesis to stop all together. When a vine has been so deprived of water it can exceed what is known as its permanent wilting point. At this point, the vine can become permanently damaged beyond recovery even if later watered. Viticulturists will carefully watch the plant for signs of severe water stress. Some of the symptoms include:

- Flaccid and wilting tendrils
- (During Flowering) Flower clusters that are dried out
- Wilting of young grape leaves followed by maturer leaves
- Chlorosis signaling that photosynthesis has stopped
- Necrosis of dying leaf tissue which leads to premature leaf fall
- Finally, the grape berries themselves start to shrivel and fall off the vine

The effectiveness of water stress is an area of continuing research in viticulture. Of particular focus is the connection between yield size and the potential benefits of water stress. Since the act of stressing the vine does contribute to reduce photosynthesis-and by extension, reduce ripening since the sugars produced by photosynthesis is needed for grape development-it is possible that a stressed vine with high yields will only produce lots of under ripe grapes.

Another interest of study is the potential impact on white grape varieties with enologists and viticulturists such as Cornelius Van Leeuwen and Catherine Peyrot Des Gachons contending that white grape varieties lose some of their aromatic qualities when subjugated to even mild forms of water stress.

Partial rootzone drying

In partial rootzone drying, half the roots are allowed to dehydrate which sends signals to the vine that is experiencing "water stress". Meanwhile the irrigated roots on the other side of vine continue to provide sufficient amounts of water so that vital functions like photosynthesis does not cease.

One irrigation technique known as partial rootzone drying (or PRD) involves "tricking" the grapevine into thinking it is undergoing water stress when it is actually receiving sufficient water supply. This is accomplished by alternating drip irrigation to where only one side of the grapevine receives water at a time. The roots on the dry side of the vine produce abscisic acid that triggers some of the vine's physiological responses to water stress - reduced shoot growth, smaller berries size, etc.

But because the vine is still receiving water on the other side the stress doesn't become so severe to where vital functions such as photosynthesis is compromised. Partial rootzone drying has been shown to significantly increase a vine's water use efficiency. While PRD is shown to slightly reduce leaf area, this is generally not a problem as overall yield is unaffected.

Criticism and environmental issues

The practice of irrigation has it share of criticism and environmental concerns. In many European wine regions the practice is banned under the belief that irrigation can be detrimental to quality wine production. However, in the early 21st century some European countries have relaxed their irrigation laws or reevaluated the issue. Of the criticisms leveled towards irrigation, the most common is that it disrupts the natural expression of terroir in the land as well as the unique characteristics that comes with vintage variation.

In regions that do not practice irrigation, the quality and styles of wines can be dramatically different from vintage to vintage depending on weather conditions and rainfall. Irrigation's contribution to the broader globalization of wine is criticized as promoting a homogenization or "standardization" of wine. Other criticisms center around the broader environmental impact of irrigation on both the ecosystem around the vineyard as well as the added strain on global water resources. While advances in drip irrigation has reduced the amount of waste water produced by irrigation, the irrigation of substantial tracts of land in areas like the San Joaquin Valley in California and the Murray-Darling Basin of Australia requires massive amounts of water from dwindling supplies. In Australia, the centuries old practice of flood irrigation used in places like the Murrumbidgee Irrigation Area caused severe environmental damages from water-logging, increase salination and raising the water tables.

In 2000, the Australian government invested over A$3.6 million into research on how to minimize the damage caused by extensive irrigation. In 2007, concerns about ecological damage to the Russian River caused government officials in California to take similar measures to cut back water supplies and promote more efficient irrigation practices.

Other uses for irrigation systems

Drip irrigation systems can also be used to distribute controlled amounts of fertilization to the vines in a process known as "fertigation". In addition to providing water for plant growth and development, irrigation systems can also be used for alternative purposes. One of the most common is the dual application of fertilizer with water in a process known as fertigation. Commonly used in drip irrigation systems, this method allows similarly regulate control over how precisely how much fertilizer and nutrients that each vine receives.

Another alternative use for sprinkler irrigation systems can occur during the threat of winter or spring time frost. When temperature drop below 32 °F (0 °C), the vine is at risk of developing frost damage that could not only ruining the upcoming years harvest but also kill the vine. One preventive measure against frost damage is to use the sprinkler irrigation system to coat the vines with a protective layer of water that freezes into ice. This layer of ice serves as insulation keeping the internal temperature of the vine from dropping below the freezing mark.

Major Irrigation Projects in India

- Nagarjunasagar (Andhra Pradesh): on the Krishna river near Nandikona village (about 44 km from Hyderabad.)
- Tungabhadra (Joint project of Andhra Pradesh and Karnatka): On the Tungabhadra river.

- Gandak (joint project of Bihar and Uttar Pradesh): Nepal also derives irrigation and power benefits from this project.
- Kosi(Bihar): A multipurpose project , which serves bihar and Nepal.
- Sone high level canal (Bihar) : an extention on sone barrage project.
- Kakrapara (Gujarat) : on the Tapti river near Kakrapara , in surat district.
- Ukal (Gujarat): A multiupurpose project, across Tapti river near Ukai village.
- Mahi (Gujarat): A two - phase project , one across the Mahi river near Wanakbori village and the other across Mahi river near Kadana.
- Sabarmati (Gujarat): A storage dam across Sabarmati river near Dhari village in Mehsana district and Wasna barrage near Ahmedabad.
- Panam (Gujarat): A masonry dam across Panam river near Keldezar village in Panchmahals district.
- Karjan(Gujarat):A masonry dam across Karjan river near Jiotgarh village in Nandoo Taluka of Bharuch district.
- Bhadra (Karnataka):A multipurpose project across the river Bhadra .
- Upper Krishna (Karnataka) : A project consisting of Narayanpur dam across the Krishna river and a dam at Almatti.
- Ghataprabha (Karnataka): A project across Ghataprabha in Belgaum and Bijapur districts.
- Malaprabha (Karnataka) : A dam across ithe Malaprabha in Belguam district.
- Tawa (Madhya Pradesh): A project the Tawa river , a tributary of the Narmada in Hoshangabad district.
- Chambal (joint project of Madhya Pradesh and rajasthan): The project comprises Gandhi Sagar dam , Rana Sagar dam and Jawahar Sagar dam.
- Mahanadi Reservoir Project (Madhya pradesh): It has three phases : (1) Ravishankar sagar project and feeder canal system for supply of water to Bhilai Steel plant and Sandur dam across Sandur village, (2) extension of Mahanadi feeder canal (3) Pairi dam.
- Hasdeo Bango Project(Madhya Pradesh): It the third phase of Hasdeo Bango Project complex and envisages construction of a masonry dam across Hasdeo river. The first and second phases have been substantially completed .

- Bargi Project (Madhya pradesh): It is a multipurpose project consisting of a masonry dam across Bargi river in the Jabalpur district and a left bank a canal.
- Bhima (Maharashtra) : Comprises two dams , one on the Pawana river near Phange in Pune district and the other across the Krishna river near Ujjaini in Sholapur district.
- Jayakwadi(Maharashtra) : A masonry spillway across the river Godavari .
- Kukadi Project (Maharashtra) : Five independent storage dams ,i.e., Yodgaon, Manikdohi, Dimba, Wadaj and Pimpalgaon jog . The canal system comprises (1) Kukadi left bank canal, (2) Dhimba left bank canal, (3) Dhimba right bank canal (4)Meena feeder and (5) Meena branch .
- Krishna Project (Maharashtra): Dhom dam near Dhom village on Krishna and Kanhar village on Varna river in Satna district.
- Upper Penganga (Mahrarhtra) : Two reservoirs on Penganga river at Isapur in Yavatmal district and the other on Rayadhu river at Sapli in Parbhani district.
- Hirakud(Orrisa): World's longest dam, is located on the Mahanadi river.
- Mahandi Delta Scheme (Orissa): The irrigation scheme will utilize releases from the Hirakud reservoir.
- Bhakra Nangal (Joint project of Haryana, Punjab and Rajasthan): India's biggest multipurpose river valley project comprises a straight gravity dam across the Satlej at Bhakra, the Nangal dam, the Nangal hydal channel, two power houses at Bhakra dam at and two power station at Ganguwal and Kotla.
- Beas (Joint venture of Haryana , Punjab , and Rajasthan) : It consists of Beas-Sutlej link and Beas dam at Pong.
- Thein Dam (Punjab): The project envisages Constroction of a dam across river Ravi and a power plant on its left bank.
- Rajasthan Canal (Rajasthan): The Project will use water released from Pong dam will provide irrigation facilities to the north western region of Rajasthan, i.e. a part of the thar desert . It consists of Rajasthan feeder canal (with the first 167 km in Punjab and Haryana and the remaining 37 km in Rajasthan) and 445 km Rajasthan main canal entirely in Rajasthan.

- Paramblkulam Allyar (Joint venture of Tamil Nadu and Kerala) : The project envisages the integrated hamessing of eight rivers , six in the Annamalai hills and two in the plains.
- Sarda Sahayak (Uttar Praesh) : The Project envisages construction of a barrage across the river Ghagra, a link channel, a barrage across river Sarda and a feeder channel involving constructions of two major aqueducts over Gomti and Sai.
- Ramganga(Uttar Pradesh): A Dam across Ramganga, a tributary of the Ganga river located in Garhwal district. The Project has, besides reducing the intensity of Floods in central and western Uttar Pradesh, provided water for the Delhi Water supply scheme.
- Left Bank Ghagra Canal (Uttar Pradesh): A link channel taking off from the left bank of Ghagra river of Girja barrage and joining with Sarju river . Also a barrage across Sarju.
- Tehri Dam (Uttar Pradesh): Earth and Rock- fill dam on Bhagirathi river in Tehri district.
- Madhya Ganga Canal (Uttar Pradesh): A barrage across ganga in Bijnor district.
- Farakka(West bengal) : The project was taken up for the preservation and maintenance of Calcutta Port and for improving the navigability of the Hooghly. It comprises a barrage across the Ganga at Farakka, a barrage at Jangipur across the Bhagirathi and a feeder channel taking off from the Ganga at Farakka and tailing at the Bhagirathi below the Jangipur barrage.
- Mayurakshi (West Bengal): An irrigation and hydro - electric Project comprises the Canada Dam.
- Kangsabati (West bengal): The project envisages construction of dams on the Kangsabati and Kumari rivers.

Damodar Valley Project(West Bengal and Bihar): A multipurpose project for the unified development of irrigation, flood control and power Generation in west Bengal and Bihar. It comprises multipurpose dams at Konar, Tilaiya, Maithon and Panchet, hydel power stations at Tilaiya, Konar, Maithon and Panchet barrage at Durgapur; and thermal power houses at Bokaro, Chandrapura and Durgapur.

2

SCOPE OF IRRIGATION ENGINEERING

Irrigation engineering is a kind of civil engineering which includes the study of controlling and harnessing the various natural sources of water. It studies the irrigation and agriculture in-depth. Flooding is the basis of agriculture and the only way by which water is supplied to the crops. Irrigation engineering is a field of engineering which would discuss the uses of irrigation in crop production and how new principles and technologies can be used and applied for irrigational practices. It is essential to study irrigation engineering as it helps in determining future irrigation expectations and impacts of irrigation in the agriculture field.

There are different types of irrigation techniques, and Irrigation engineering includes the study of the design of different irrigation systems and analyzes the efficiency of each of the system and list their advantages and disadvantages. People studying irrigation engineering will also consider the different cases to understand when irrigation is necessary and when it is most important.

The process of artificial utilization of water to the soil for the growth of agricultural crops is named as irrigation. It is basically a study of arranging and planning a water supply framework for the agricultural land to shield the harvests from terrible impact of drought or minimum rainfall. Irrigation engineering contains the investigation and the plan of works for association with river or waterway control, drainage of water logged zones and generates the hydroelectric power.

Irrigation Engineering is very important for designing and planning of cost-effective and efficient irrigation systems. There are various benefits as well as risks of the irrigation system. For example, it increases the groundwater level and aid to the cultivation and also protects a country from a famine situation. However, there are some ill-effects too of irrigation such as it creates a breeding place for the mosquitoes. Also, valuable land may be submerged because of the construction of dams, and barrages. The benefits are quite more than the ill-effects of the irrigation system.

Irrigation projects are somewhat complicated and expensive. Therefore, a trained professional is required to understand and design the plan which is why there is a need for an irrigation engineer who can handle and manage all these tasks. The primary responsibility of an irrigation engineer is not only related to the application of the water. Instead, it deals with various other aspects.

It deals with the designing and planning of the irrigation systems.

It deals with the problems that may arise in the watershed or the agricultural fields.

The study of irrigation engineering deals with the rivers and the hydrology.

It is the study of water, soil, and crop relationship.

It deals with the designs and structure of dams, canals, other hydraulic and irrigation structures.

The scope of Irrigation Engineering

The scope of irrigation engineering is a vast practice in civil engineering field. Irrigation engineering manages the arranging, planning and keeping up with irrigation frameworks for the vehicle and dispersion of water to private and farming destinations. Here, it is discussed about the scope of irrigation engineering, benefits of irrigation engineering, ill effects of irrigation engineering and necessity of irrigation engineering.

The scope of irrigation engineering is a vast practice in civil engineering field. Irrigation engineering manages the arranging, planning and keeping up with irrigation frameworks for the vehicle and dispersion of water to private and farming destinations. The works involved in irrigation engineering may oversee the installation of pipelines and sprinklers or be involved in directing water from dams, canals and rivers. The work is done in the field to evaluate terrain, soil and climatic characteristics to optimize the use of water for lawns and agricultural crops. Computers are often employed for modelling techniques to plan blueprints of irrigation systems and present these to sub-contractors for installation technicians.

Irrigation engineering addresses ongoing issues of drought and national water shortages, focus on developing irrigation systems that prioritize water sustainability.

Irrigation engineering may be involved in researching water recycling methods and developing new ways of minimizing water seepage and evaporation loss.

Irrigation engineering is regularly utilized in rural ventures like cotton, dairy and agriculture. Other works involved the residential and business sectors in the construction of irrigation systems for home-owners and office buildings.

Here, it is discussed about the scope of irrigation engineering, benefits of irrigation engineering, ill effects of irrigation engineering and scope and necessity of irrigation engineering.

Planning and Designing.

Construction.

Management of Irrigation Systems.

Operations.

Lead irrigation projects.

Provide better facilities and methods for irrigation.

The main job of the irrigation engineer is to find the efficient and cost-effective ways of an irrigation system which costs less and yield better results.

To be an irrigation engineer, you should have a basic understanding of basic science and engineering, physics, plants, soil, water and their relationship, hydrology, and hydraulics.

NECESSITY OF IRRIGATION ENGINEERING

There are many scope and necessity of irrigation engineering. All through the yield time frame sufficient amounts of water is needed close to the root zone of the plants for their development. At times, during the crop period, the rainfall may not be satisfactory to satisfy the water requirement of the crops. The intensity of rainfall is basically unsure and outside the ability to control and it may not be all around dispersed all through the crop season or the culturable area. Along these lines, water system turns out to be totally important to satisfy the water prerequisite of the crops.

The necessity of irrigation engineering can be summarized in the following points:

1. Less rainfall: When the total rainfall is less they are needed for the crops; hence artificial supply is necessary. In such a case water system works might be developed at a spot where more water is accessible and then to pass on the water to the area where there is insufficiency of water.
2. Uneven distribution of rainfall: When the rainfall is not evenly distributed during the crop period, the irrigation is extremely necessary. By the collection of water during the abundance rainfall period, water might be provided to the crop during the period when there might be no rainfall.
3. Improvement of perennial crops: Some perennial crops like sugarcane, cotton etc. require water all through the significant piece of the year. But the rainfall may satisfy the water necessity in the rainy season only. So, for the remaining part of the year irrigations become necessary.
4. Development of agriculture in dessert area: In desert areas where the rainfall is very scanty, irrigation is required for the development of agriculture.
5. Growing a number of crops during a year: The rainfall in an area may be sufficient to raise only one type of crop during the rainy season for which no irrigation may be required. However, with the provision of irrigation facilities in that area, crops can be raised in other seasons also.

Benefits and Ill Effects of Irrigation Engineering

An irrigation engineering has both the benefits and ill effects of irrigation engineering. Some of the major benefits of irrigation engineering and ill effects of irrigation engineering are mentioned as below.

Advantages of Irrigation Engineering

Here it is described about some benefits of irrigation engineering:

1. Yield of crops: In the time of low rainfall or drought, the yield of crop might be expanded by the water system framework.

2. Protection from famine: The food creation of a nation can be improved by guaranteeing the development of crops by availing the irrigation facilities. A country is prevented from famine situation by irrigation engineering.
3. Prosperity of farmers: When the supply of irrigation water is assured, the farmers can grow two or more crops in a year on the same land which is a proper benefits of irrigation engineering. Thus, the farmers may earn more money and improve their living standards.
4. Improvement of cash cops: Irrigation helps to improve the cultivation of cash crops like tobacco, cotton, sugarcane, coffee, tea etc.
5. Domestic and industrial water supply: The irrigation canals may be utilized for domestic and industrial water supply. Some of the water system reservoirs likewise supply water to close by rural and urban areas. The canals also provide facilities for bathing, boating and other recreations.
6. Development of fishery: The reservoirs and the canals can be utilized for development of fishery projects.

Disadvantages of Irrigation Engineering

Here it is described about some ill effects of irrigation engineering:

1. Breeding places for mosquitos: Due to access application of water and due to leakage of water, ponds and depressions get filled up with water and create breeding places for mosquitos. Also if the canal is leaky, mosquitos breed all along the canal and spread malarial conditions.
2. Water logging: Due to excessive seepage of water through the bed and banks of the canal, the water table in the surrounding area may be raised which may constantly saturate the root zone of the crops and the whole area becomes water logged.
3. Dampness in weather: The areas which are already damp and cold, becomes damper and colder due to irrigation. Because of dampness, the individuals living around the region may experience the ill effects of cold, cough and other such infections originating from dampness.
4. Loss of valuable land: Important land may be lowered due to developments of reservoirs and damps and it might likewise be lost while building irrigation canals.

3

HYDROLOGY AND IRRIGATION ENGINEERING

Hydrology in civil engineering may be understood as the branch of engineering that deals with the study of water resources. It is also commonly referred to as water resources engineering. Precisely, engineering hydrology may be defined as the multidisciplinary specialization of civil engineering that deals with the occurrence, circulation, and distribution of water on Earth.

Hydrology is the science of water which deals with the spatial and temporal characteristics of the earth's water. In other words, hydrology is the study of water in all its phases, including the occurrence, movement, distribution, circulation, storage, exploration, development and management.

Hydrology (from Ancient Greek ?d?? (húdor) 'water', and -????a (-logía) 'study of') is the scientific study of the movement, distribution, and management of water on Earth and other planets, including the water cycle, water resources, and environmental watershed sustainability. A practitioner of hydrology is called a hydrologist. Hydrologists are scientists studying earth or environmental science, civil or environmental engineering, and physical geography. Using various analytical methods and scientific techniques, they collect and analyze data to help solve water related problems such as environmental preservation, natural disasters, and water management.

The atmospheric air always contains moisture. Under favourable conditions, this moisture gets condensed and water drops are formed. These water drops ultimately fall on the earth's surface as precipitation in the form of rain, snow, fog, drizzle, dew, hail (frozen rain drops) and sleet (snow mixed with rain).

Rainfall is the water that reaches the surface of the earth from the atmosphere due to the condensation of the atmospheric vapour. Part of the natural precipitation (i.e., rainfall) is transported on the earth's surface through natural surface channels which are known as streams or rivers.

This is known as runoff which includes:

(i) Surface runoff or overland flow received in a stream immediately after rain,

(ii) Interflow, i.e. the portion of soil moisture which flows through the upper soil layers and joins the stream, and

(iii) Delayed runoff which includes part of rain water which enters the stream after flowing through deeper portions of the ground, and that part of the runoff which has been temporarily stored in natural lakes or as snow cover.

Overland flow and interflow together are termed direct runoff, while the ground water flow that enters a stream is termed base flow (or base runoff). Thus, runoff is the total quantity of water received by a stream from its drainage basin (or watershed or catchment). A drainage basin of a stream is defined as that part of the earth's surface which diverts all its runoff to the same stream (i.e., drainage outlet).

Characteristics of a drainage basin such as size, shape, surface characteristics, topography and geology and those of rain (like its intensity and duration) considerably affect the runoff. In general, a larger catchment would yield higher runoff. A fan-shaped catchment produces larger runoff compared to that produced by fern-shaped catchment.

SHAPES OF CATCHMENTS

Similarly, a catchment with vegetal cover yields lesser runoff compared to that yielded by a catchment with paved surface or no vegetal cover. A steeply sloping catchment will result in higher runoff. Again, a catchment with pervious soil will yield lesser runoff due to larger infiltration into the ground. Also, the runoff increases with the increase in the intensity of rainfall and decrease in the duration of rainfall.

The runoff (or surface runoff) ultimately meets the sea. Part of the runoff may percolate deep into the ground and become part of ground water just as part of the natural precipitation that falls on the ground may percolate directly deep into the ground without contributing to the runoff. Part of this ground water may join runoff later as delayed runoff.

For the purpose of providing irrigation facilities, one has to plan, construct and operate different types of irrigation (or water resources) structures, such as clams, barrages, irrigation channels, canal structures, etc. Besides, bridges have to be constructed across rivers, and several types of measures have to be taken to minimise loss of life and property due to floods in rivers.

Likewise, hydropower projects are constructed for the benefit of mankind. All such projects require knowledge about the stream flow and its behaviour. It is this (i.e., the behaviour of a stream) and related aspects which make study of hydrology useful and important for irrigation engineers.

History of Engineering Hydrology

Engineering hydrology has been an important study for a very long time.

Ever since the dawn of civilization, various hydraulic structures have been constructed by the use of hydrological tools.

For example, during the Mesopotamian civilization dams were constructed for protecting the Mesopotamian towns from floods. Numerous aqueducts were built by the Greeks and Romans.

Irrigation activities were evident since the history of Chinese development.

The development and advancement of engineering hydrology began in the first century BC.

Robert Elmer Horton is regarded as the "Father of Modern Engineering Hydrology". He was an American civil engineer and soil scientist.

Importance of Hydrology in Civil Engineering

Engineering Hydrology and its applications are inevitable for planning and building hydraulic structures.

It is important for solving the various existing water-related problems such as problems regarding the quantity of water, control, and mitigation of flood, irrigation, etc.

Some of the importance of engineering hydrology can be listed as follows:

1. It is necessary for determining the maximum probable flood at the proposed construction site. For example, dam construction.
2. It facilitates engineers and hydrologists for establishing the relation between surface water of catchment and underground water resources.
3. It enables one to determine the flow over various hydraulic structures such as spillways, highway culverts, urban storm drainage systems, etc.

4. It is important for studying the on-site drainage and seepage condition before the commencement of any engineering construction.
5. The nature of variations of water flow, rainfall pattern, etc can be obtained from the use of engineering hydrology applications.
6. It is necessary for determining the reservoir capacity to assure an adequate supply of water for domestic and other purposes.

Application Area of Hydrology in Civil Engineering

The main application areas in the field of civil engineering can be listed as follows:

1. Irrigation Projects and Schemes.
2. Hydroelectric Power Generation Projects and Dam Constructions.
3. Water Supply Projects.
4. Disaster Prevention or Flood Control Projects.
5. Scope of Hydrology in Civil Engineering

The main scope of engineering hydrology is briefly described as follows:

1. Determination of Maximum Probable Flood:

The maximum probable flood at a proposed site can be determined by the techniques of engineering hydrology.

It is necessary to determine the frequency and occurrence of floods for designing and building hydraulic structures such as dams, flood control structures, and reservoirs.

2. Determination of Maximum Intensity of Storm:

The determination of the maximum intensity of the storm is necessary to determine the drainage conditions at the site and to design the needed drainage systems.

3. Determination of Water Yield of Basin:

For the design and construction of various structures such as municipal water tanks, reservoirs, inland navigation, etc; it is necessary to determine the yield of the basin i.e. the occurrence, frequency, and quantity of water that can be obtained from the basin.

4. Ground Water Development Study:

Engineering hydrology facilitates the study of groundwater and its development.

Water Cycle

The water cycle, also known as the hydrologic cycle or H2O cycle, describes the continuous movement of water on, above and below the surface of the Earth. Water can change states among liquid, vapor, and ice at various places in the water cycle. Although the balance of water on Earth remains fairly constant over time, individual water molecules can come and go, in and out of the atmosphere.

The water moves from one reservoir to another, such as from river to ocean, or from the ocean to the atmosphere, by the physical processes of evaporation, condensation, precipitation, infiltration, runoff, and subsurface flow. In so doing, the water goes through different phases: liquid, solid, and gas.

The hydrologic cycle also involves the exchange of heat energy, which leads to temperature changes. For instance, in the process of evaporation, water takes up energy from the surroundings and cools the environment. Conversely, in the process of condensation, water releases energy to its surroundings, warming the environment.

The water cycle figures significantly in the maintenance of life and ecosystems on Earth. Even as water in each reservoir plays an important role, the water cycle brings added significance to the presence of water on our planet. By transferring water from one reservoir to another, the water cycle purifies water, replenishes the land with freshwater, and transports minerals to different parts of the globe. It is also involved in reshaping the geological features of the Earth, through such processes as erosion and sedimentation. In addition, as the water cycle involves heat exchange, it exerts an influence on climate as well.

The sun, which drives the water cycle, heats water in oceans and seas. Water evaporates as water vapor into the air. Ice and snow can sublimate directly into water vapor. Evapotranspiration is water transpired from plants and evaporated from the soil. Rising air currents take the vapor up into the atmosphere where cooler temperatures cause it to condense into clouds. Air currents move water vapor around the globe, cloud particles collide, grow, and fall out of the sky as precipitation.

Some precipitation falls as snow or hail, and can accumulate as ice caps and glaciers, which can store frozen water for thousands of years. Snowpacks can thaw and melt, and the melted water flows over land as snowmelt. Most water falls back into the oceans or onto land as rain, where the water flows over the ground as surface runoff. A portion of runoff enters rivers in valleys in the landscape, with streamflow moving water towards the oceans. Runoff and groundwater are stored as freshwater in lakes.

Not all runoff flows into rivers, much of it soaks into the ground as infiltration. Some water infiltrates deep into the ground and replenishes aquifers, which store freshwater for long periods of time. Some infiltration stays close to the land surface and can seep back into surface-water bodies (and the ocean) as groundwater discharge. Some groundwater finds openings in the land surface and comes out as freshwater springs. Over time, the water returns to the ocean, where our water cycle started.

Different Processes

Precipitation

Condensed water vapor that falls to the Earth's surface . Most precipitation occurs as rain, but also includes snow, hail, fog drip, graupel, and sleet. Approximately 505,000 km3 (121,000 cu mi) of water falls as precipitation each year, 398,000 km3 (95,000 cu mi) of it over the oceans.

Canopy interception

The precipitation that is intercepted by plant foliage and eventually evaporates back to the atmosphere rather than falling to the ground.

Snowmelt

The runoff produced by melting snow.

Runoff

The variety of ways by which water moves across the land. This includes both surface runoff and channel runoff. As it flows, the water may seep into the ground, evaporate into the air, become stored in lakes or reservoirs, or be extracted for agricultural or other human uses.

Infiltration

The flow of water from the ground surface into the ground. Once infiltrated, the water becomes soil moisture or groundwater.

Subsurface Flow

The flow of water underground, in the vadose zone and aquifers. Subsurface water may return to the surface (e.g. as a spring or by being pumped) or eventually seep into the oceans. Water returns to the land surface at lower elevation than where it infiltrated, under the force of gravity or gravity induced pressures. Groundwater tends to move slowly, and is replenished slowly, so it can remain in aquifers for thousands of years.

Evaporation

The transformation of water from liquid to gas phases as it moves from the ground or bodies of water into the overlying atmosphere. The source of energy for evaporation is primarily solar radiation. Evaporation often implicitly includes transpiration from plants, though together they are specifically referred to as evapotranspiration. Total annual evapotranspiration amounts to approximately 505,000 km^3 (121,000 cu mi) of water, 434,000 km^3 (104,000 cu mi) of which evaporates from the oceans.

Sublimation

The state change directly from solid water (snow or ice) to water vapor.

Advection

The movement of water - in solid, liquid, or vapor states - through the atmosphere. Without advection, water that evaporated over the oceans could not precipitate over land.

Condensation

The transformation of water vapor to liquid water droplets in the air, creating clouds and fog.

Transpiration

The release of water vapor from plants and soil into the air. Water vapor is a gas that cannot be seen.

The residence time of a reservoir within the hydrologic cycle is the average time a water molecule will spend in that reservoir (see adjacent table). It is a measure of the average age of the water in that reservoir.

Groundwater can spend over 10,000 years beneath Earth's surface before leaving. Particularly old groundwater is called fossil water. Water stored in the soil remains there very briefly, because it is spread thinly across the Earth, and is readily lost by evaporation, transpiration, stream flow, or groundwater recharge. After evaporating, the residence time in the atmosphere is about 9 days before condensing and falling to the Earth as precipitation.

The major ice sheets - Antarctica and Greenland - store ice for very long periods. Ice from Antarctica has been reliably dated to 800,000 years before present, though the average residence time is shorter.

In hydrology, residence times can be estimated in two ways. The more common method relies on the principle of conservation of mass and assumes the amount of water in a given reservoir is roughly constant. With this method, residence times are estimated by dividing the volume of the reservoir by the rate by which water either enters or exits the reservoir.

Conceptually, this is equivalent to timing how long it would take the reservoir to become filled from empty if no water were to leave (or how long it would take the reservoir to empty from full if no water were to enter). An alternative method to estimate residence times, which is gaining in popularity for dating groundwater, is the use of isotopic techniques. This is done in the subfield of isotope hydrology.

CHANGES OVER TIME

The water cycle describes the processes that drive the movement of water throughout the hydrosphere. However, much more water is "in storage" for long periods of time than is actually moving through the cycle. The storehouses for the vast majority of all water on Earth are the oceans. It is estimated that of the 332,500,000 mi^3 (1,386,000,000 km^3) of the world's water supply, about 321,000,000 mi^3 (1,338,000,000 km^3) is stored in oceans, or about 95%. It is also estimated that the oceans supply about 90% of the evaporated water that goes into the water cycle.

During colder climatic periods more ice caps and glaciers form, and enough of the global water supply accumulates as ice to lessen the amounts in other parts of the water cycle. The reverse is true during warm periods. During the last ice age glaciers covered almost one-third of Earth's land mass, with the result being that the oceans were about 400 ft (122 m) lower than today. During the last global "warm spell," about 125,000 years ago, the seas were about 18 ft (5.5 m) higher than they are now. About three million years ago the oceans could have been up to 165 ft (50 m) higher.

The scientific consensus expressed in the 2007 Intergovernmental Panel on Climate Change (IPCC) Summary for Policymakers is for the water cycle to continue to intensify throughout the 21st century, though this does not mean that precipitation will increase in all regions. In subtropical land areas - places that are already relatively dry - precipitation is projected to decrease during the 21st century, increasing the probability of drought.

The drying is projected to be strongest near the poleward margins of the subtropics (for example, the Mediterranean Basin, South Africa, southern Australia, and the Southwestern United States). Annual precipitation amounts are expected to increase in near-equatorial regions that tend to be wet in the present climate, and also at high latitudes. These large-scale patterns are present in nearly all of the climate model simulations conducted at several international research centers as part of the 4th Assessment of the IPCC. Glacial retreat is also an example of a changing water cycle, where the supply of water to glaciers from precipitation cannot keep up with the loss of water from melting and sublimation. Glacial retreat since 1850 has been extensive.

Human activities that alter the water cycle include:

- Agriculture
- Industry
- Alteration of the chemical composition of the atmosphere
- Construction of dams
- Deforestation and afforestation
- Removal of groundwater from wells
- Water abstraction from rivers
- Urbanization

Effects on climate

The water cycle is powered from solar energy. 86% of the global evaporation occurs from the oceans, reducing their temperature by evaporative cooling. Without the cooling, the effect of evaporation on the greenhouse effect would lead to a much higher surface temperature of 67 °C (153 °F), and a warmer planet.

Aquifer drawdown or overdrafting and the pumping of fossil water increases the total amount of water in the hydrosphere that is subject to transpiration and evaporation thereby causing accretion in water vapour and cloud cover which are the primary absorbers of infrared radiation in the Earth's atmosphere. Adding water to the system has a forcing effect on the whole earth system, an accurate estimate of which hydrogeological fact is yet to be quantified.

Effects on biogeochemical cycling

While the water cycle is itself a biogeochemical cycle, flow of water over and beneath the Earth is a key component of the cycling of other biogeochemicals. Runoff is responsible for almost all of the transport of eroded sediment and phosphorus from land to waterbodies. The salinity of the oceans is derived from erosion and transport of dissolved salts from the land. Cultural eutrophication of lakes is primarily due to phosphorus, applied in excess to agricultural fields in fertilizers, and then transported overland and down rivers. Both runoff and groundwater flow play significant roles in transporting nitrogen from the land to waterbodies. The dead zone at the outlet of the Mississippi River is a consequence of nitrates from fertilizer being carried off agricultural fields and funnelled down the river system to the Gulf of Mexico. Runoff also plays a part in the carbon cycle, again through the transport of eroded rock and soil.

The hydrodynamic wind within the upper portion of a planet's atmosphere allows light chemical elements such as Hydrogen to move up to the exobase, the lower limit of the exosphere, where the gases can then reach escape velocity, entering outer space without impacting other particles of gas. This type of gas loss from a planet into space is known as planetary wind. Planets with hot lower atmospheres could result in humid upper atmospheres that accelerate the loss of hydrogen.

Groundwater

Groundwater is water located beneath the ground surface in soil pore spaces and in the fractures of rock formations. A unit of rock or an unconsolidated deposit is called an aquifer when it can yield a usable quantity of water. The depth at which soil pore spaces or fractures and voids in rock become completely saturated with water is called the water table. Groundwater is recharged from, and eventually flows to, the surface naturally; natural discharge often occurs at springs and seeps, and can form oases or wetlands. Groundwater is also often withdrawn for agricultural, municipal and industrial use by constructing and operating extraction wells. The study of the distribution and movement of groundwater is hydrogeology, also called groundwater hydrology.

Typically, groundwater is thought of as liquid water flowing through shallow aquifers, but technically it can also include soil moisture, permafrost (frozen soil), immobile water in very low permeability bedrock, and deep geothermal or oil formation water. Groundwater is hypothesized to provide lubrication that can possibly influence the movement of faults. It is likely that much of the Earth's subsurface contains some water, which may be mixed with other fluids in some instances. Groundwater may not be confined only to the Earth. The formation of some of the landforms observed on Mars may have been influenced by groundwater. There is also evidence that liquid water may also exist in the subsurface of Jupiter's moon Europa.

Aquifers

An aquifer is a layer of porous substrate that contains and transmits groundwater. When water can flow directly between the surface and the saturated zone of an aquifer, the aquifer is unconfined. The deeper parts of unconfined aquifers are usually more saturated since gravity causes water to flow downward.

The upper level of this saturated layer of an unconfined aquifer is called the water table or phreatic surface. Below the water table, where generally all pore spaces are saturated with water is the phreatic zone.

Substrate with low porosity that permits limited transmission of groundwater is known as an aquitard. An aquiclude is a substrate with porosity that is so low it is virtually impermeable to groundwater.

A confined aquifer is an aquifer that is overlain by a relatively impermeable layer of rock or substrate such as an aquiclude or aquitard. If a confined aquifer follows a downward grade from its recharge zone, groundwater can become pressurized as it flows. This can create artesian wells that flow freely without the need of a pump and rise to a higher elevation than the static water table at the above, unconfined, aquifer.

The characteristics of aquifers vary with the geology and structure of the substrate and topography in which they occur. Generally, the more productive aquifers occur in sedimentary geologic formations. By comparison, weathered and fractured crystalline rocks yield smaller quantities of groundwater in many environments. Unconsolidated to poorly cemented alluvial materials that have accumulated as valley-filling sediments in major river valleys and geologically subsiding structural basins are included among the most productive sources of groundwater.

The high specific heat capacity of water and the insulating effect of soil and rock can mitigate the effects of climate and maintain groundwater at a relatively steady temperature. In some places where groundwater temperatures are maintained by this effect at about 50°F/10°C, groundwater can be used for controlling the temperature inside structures at the surface. For example, during hot weather relatively cool groundwater can be pumped through radiators in a home and then returned to the ground in another well. During cold seasons, because it is relatively warm, the water can be used in the same way as a source of heat for heat pumps that is much more efficient than using air.

Relative groundwater travel times

Groundwater makes up about twenty percent of the world's fresh water supply, which is about 0.61% of the entire world's water, including oceans and permanent ice. Global groundwater storage is roughly equal to the total amount of freshwater stored in the snow and ice pack, including the north and south poles. This makes it an important resource which can act as a natural storage that can buffer against shortages of surface water, as in during times of drought.

Groundwater is naturally replenished by surface water from precipitation, streams, and rivers when this recharge reaches the water table.

Groundwater can be a long-term 'reservoir' of the natural water cycle (with residence times from days to millennia), as opposed to short-term water reservoirs like the atmosphere and fresh surface water (which have residence times from minutes to years). The figure shows how deep groundwater (which is quite distant from the surface recharge) can take a very long time to complete its natural cycle.

The Great Artesian Basin in central and eastern Australia is one of the largest confined aquifer systems in the world, extending for almost 2 million km2. By analysing the trace elements in water sourced from deep underground, hydrogeologists have been able to determine that water extracted from these aquifers can be more than 1 million years old.

By comparing the age of groundwater obtained from different parts of the Great Artesian Basin, hydrogeologists have found it increases in age across the basin. Where water recharges the aquifers along the Eastern Divide, ages are young. As groundwater flows westward across the continent, it increases in age, with the oldest groundwater occurring in the western parts. This means that in order to have travelled almost 1000 km from the source of recharge in 1 million years, the groundwater flowing through the Great Artesian Basin travels at an average rate of about 1 metre per year.

Issues

Overview

Certain problems have beset the use of groundwater around the world. Just as river waters have been over-used and polluted in many parts of the world, so too have aquifers. The big difference is that aquifers are out of sight. The other major problem is that water management agencies, when calculating the 'sustainable yield' of aquifer and river water, have often counted the same water twice, once in the aquifer, and once in its connected river. This problem, although understood for centuries, has persisted, partly through inertia within government agencies. In Australia, for example, prior to the statutory reforms initiated by the Council of Australian Governments water reform framework in the 1990s, many Australian States managed groundwater and surface water through separate government agencies, an approach beset by rivalry and poor communication.

The time lags inherent in the dynamic response of groundwater to development have generally been ignored by water management agencies, decades after scientific understanding of the issue was consolidated. In brief, the effects of groundwater overdraft (although undeniably real) may take decades or centuries to manifest themselves. In a classic study in 1982, Bredehoeft and colleagues modelled a situation where groundwater extraction in an intermontane basin withdrew the entire annual recharge, leaving 'nothing' for the natural groundwater-dependent vegetation community. Even when the borefield was situated close to the vegetation, 30% of the original vegetation demand could still be met by the lag inherent in the system after 100 years. By year 500 this had reduced to 0%, signalling complete death of the groundwater-dependent vegetation. The science has been available to make these calculations for decades; however water management agencies have generally ignored effects which will appear outside the rough timeframe of political elections (3 to 5 years). Marios Sophocleous argued strongly that management agencies must define and use appropriate timeframes in groundwater planning. This will mean calculating groundwater withdrawal permits based on predicted effects decades, sometimes centuries in the future.

As water moves through the landscape it collects soluble salts, mainly sodium chloride. Where such water enters the atmosphere through evapotranspiration, these salts are left behind. In irrigation districts, poor drainage of soils and surface aquifers can result in water tables coming to the surface in low-lying areas. Major land degradation problems of soil salinity and waterlogging result, combined with increasing levels of salt in surface waters. As a consequence, major damage has occurred to local economies and environments.

Four important effects are worthy of brief mention. First, flood mitigation schemes, intended to protect infrastructure built on floodplains, have had the unintended consequence of reducing aquifer recharge associated with natural flooding. Second, prolonged depletion of groundwater in extensive aquifers can result in land subsidence, with associated infrastructure damage - as well as (thirdly) saline intrusion. Fourth, draining acid sulphate soils, often found in low-lying coastal plains, can result in acidification and pollution of formerly freshwater and estuarine streams.

Another cause for concern is that groundwater drawdown from over-allocated aquifers has the potential to cause severe damage to both terrestrial and aquatic ecosystems - in some cases very conspicuously but in others quite imperceptibly because of the extended period over which the damage occurs.

Overdraft

Groundwater is a highly useful and often abundant resource. However, over-use, or overdraft, can cause major problems to human users and to the environment. The most evident problem (as far as human groundwater use is concerned) is a lowering of the water table beyond the reach of existing wells. Wells must consequently be deepened to reach the groundwater; in some places (e.g., California, Texas and India) the water table has dropped hundreds of feet because of extensive well pumping. In the Punjab region of India, for example, groundwater levels have dropped 10 meters since 1979, and the rate of depletion is accelerating. A lowered water table may, in turn, cause other problems such as groundwater-related subsidence and saltwater intrusion.

Groundwater is also ecologically important. The importance of groundwater to ecosystems is often overlooked, even by freshwater biologists and ecologists. Groundwaters sustain rivers, wetlands and lakes, as well as subterranean ecosystems within karst or alluvial aquifers.

Not all ecosystems need groundwater, of course. Some terrestrial ecosystems - for example, those of the open deserts and similar arid environments - exist on irregular rainfall and the moisture it delivers to the soil, supplemented by moisture in the air. While there are other terrestrial ecosystems in more hospitable environments where groundwater plays no central role, groundwater is in fact fundamental to many of the world's major ecosystems. Water flows between groundwaters and surface waters. Most rivers, lakes and wetlands are fed by, and (at other places or times) feed groundwater, to varying degrees. Groundwater feeds soil moisture through percolation, and many terrestrial vegetation communities depend directly on either groundwater or the percolated soil moisture above the aquifer for at least part of each year. Hyporheic zones (the mixing zone of streamwater and groundwater) and riparian zones are examples of ecotones largely or totally dependent on groundwater.

Aquifer drawdown or overdrafting and the pumping of fossil water increases the total amount of water within the hydrosphere subject to transpiration and evaporation processes, thereby causing accretion in water vapour and cloud cover, the primary absorbers of infrared radiation in the Earth's atmosphere. Adding water to the system has a forcing effect on the whole earth system, an accurate estimate of which hydrogeological fact is yet to be quantified.

Subsidence

Subsidence occurs when too much water is pumped out from underground, deflating the space below the above-surface, and thus causing the ground to actually collapse. The result can look like craters on plots of land. This occurs because in its natural equilibrium state, the hydraulic pressure of groundwater in the pore spaces of the aquifer and the aquitard supports some of the weight of the overlying sediments. When groundwater is removed from aquifers by excessive pumping, pore pressures in the aquifer drop and compression of the aquifer may occur. This compression may be partially recoverable if pressures rebound, but much of it is not. When the aquifer gets compressed it may cause land subsidence, a drop in the ground surface. The city of New Orleans, Louisiana, is actually below sea level today, and its subsidence is partly caused by removal of groundwater from the various aquifer/aquitard systems beneath it. In the first half of the 20th century, the city of San Jose, California, dropped 13 feet from land subsidence caused by overpumping; this subsidence has been halted with improved groundwater management.

Seawater intrusion

Generally, in very humid or undeveloped regions, the shape of the water table mimics the slope of the surface. The recharge zone of an aquifer near the seacoast is likely to be inland, often at considerable distance. In these coastal areas, a lowered water table may induce sea water to reverse the flow toward the land. Sea water moving inland is called a saltwater intrusion. Alternatively, salt from mineral beds may leach into the groundwater of its own accord.

Mining

Sometimes the water movement from the recharge zone to the place where it is withdrawn may take centuries (see figure above). When the usage of water is greater than the recharge, it is referred to as mining water (the water is often called fossil water because of its geologic age). Under those circumstances it is not a renewable resource.

Pollution

Water pollution of groundwater, from pollutants released to the ground that can work their way down into groundwater, can create a contaminant plume within an aquifer. Movement of water and dispersion within the aquifer spreads the pollutant over a wider area, its advancing boundary often called a plume edge, which can then intersect with groundwater wells or daylight into surface water such as seeps and springs, making the water supplies unsafe for humans and wildlife. The interaction of groundwater contamination with surface waters is analyzed by use of hydrology transport models.

The stratigraphy of the area plays an important role in the transport of these pollutants. An area can have layers of sandy soil, fractured bedrock, clay, or hardpan. Areas of karst topography on limestone bedrock are sometimes vulnerable to surface pollution from groundwater. Earthquake faults can also be entry routes for downward contaminant entry. Water table conditions are of great importance for drinking water supplies, agricultural irrigation, waste disposal (including nuclear waste), wildlife habitat, and other ecological issues. In the US, upon commercial real estate property transactions both groundwater and soil are the subjects of scrutiny, with a Phase I Environmental Site Assessment normally being prepared to investigate and disclose potential pollution issues. In the San Fernando Valley of California, Real estate contracts for property transfer below the Santa Susana Field Laboratory (SSFL) and eastward have clauses releasing the seller from liability for groundwater contamination consequences from existing or future water pollution of the Valley Aquifer.

Love Canal was one of the most widely known examples of groundwater pollution. In 1978, residents of the Love Canal neighbourhood in upstate New York noticed high rates of cancer and an alarming number of birth defects. This was eventually traced to organic solvents and dioxins from an industrial landfill that the neighbourhood had been built over and around, which had then infiltrated into the water supply and evaporated in basements to further contaminate the air. Eight hundred families were reimbursed for their homes and moved, after extensive legal battles and media coverage.

Another example of widespread groundwater pollution is in the Ganges Plain of northern India and Bangladesh where severe contamination of groundwater by naturally occurring arsenic affects 25% of water wells in the shallower of two regional aquifers. The pollution occurs because aquifer sediments contain organic matter (dead plant material) that generates anaerobic (an environment without oxygen) conditions in the aquifer. These conditions result in the microbial dissolution of iron oxides in the sediment and thus the release of the arsenic, normally strongly bound to iron oxides, into the water. As a consequence, arsenic-rich groundwater is often iron-rich, although secondary processes often obscure the association of dissolved arsenic and dissolved iron

WATER WELL

Water well is an excavation or structure created in the ground by digging, driving, boring or drilling to access groundwater in underground aquifers. The well water is drawn by an electric submersible pump, a trash pump, a vertical turbine pump, a handpump or a mechanical pump (e.g. from a water-pumping windmill). It can also be drawn up using containers, such as buckets, that are raised mechanically or by hand.

Wells can vary greatly in depth, water volume and water quality. Well water typically contains more minerals in solution than surface water and may require treatment to soften the water by removing minerals such as arsenic, iron and manganese.

Types of water wells

Dug wells

Until recent centuries, all artificial wells were pumpless hand-dug wells of varying degrees of formality, and they remain a very important source of potable water in some rural, developing areas where they are routinely dug and used today. Their indispensability has produced a number of literary references, literal and figurative, to them, including the Christian Bible story of Jesus meeting a woman at Jacob's well (John 4:6) and the "Ding Dong Bell" nursery rhyme about a cat in a well.

Hand-dug wells are excavations with diameters large enough to accommodate one or more men with shovels digging down to below the water table. They can be lined with laid stones or brick; extending this lining upwards above the ground surface into a wall around the well serves to reduce both contamination and injuries by falling into the well. A more modern method called caissoning uses reinforced concrete or plain concrete pre-cast well rings that are lowered into the hole. A well-digging team digs under a cutting ring and the well column slowly sinks into the aquifer, whilst protecting the team from collapse of the well bore.

Hand dug wells provide a cheap and low-tech solution to accessing groundwater in rural locations in developing countries, and may be built with a high degree of community participation, or by local entrepreneurs who specialize in hand-dug wells. Hand dug wells have been successfully excavated to 60m. Hand dug wells are inexpensive and low tech (compared to drilling) as they use mostly hand labour for construction.

Hand dug wells have low operational and maintenance costs, in part because water can be extracted by hand bailing, without a pump. Hand dug wells can be easily deepened, which may be necessary if the ground water level drops, by telescoping the lining further down into the aquifer. The yield of existing hand dug wells may be improved by deepening or introducing vertical tunnels or perforated pipes.

Drawbacks to hand-dug wells are numerous. It can be impractical to hand dig wells in areas where hard rock is present, and they can be time-consuming to dig and line even in favorable areas. Because they exploit shallow aquifers, the well may be susceptible to yield fluctuations and possible contamination from surface water, including sewage.

Hand dug well construction generally requires the use of a trained well construction team, and the capital investment for equipment such as concrete ring moulds, heavy lifting equipment, well shaft formwork, motorized de-watering pumps, and fuel can be large for people in developing countries. Construction of hand dug wells can be dangerous due to collapse of the well bore, falling objects and asphyxiation, including from dewatering pump exhaust fumes. Woodingdean well, dug between 1858 and 1862, is claimed to be the world's deepest hand-dug well at 1,285 feet.

Driven wells

Driven wells may be very simply created in unconsolidated material with a "well point", which consists of a hardened drive point and a screen (perforated pipe). The point is simply hammered into the ground, usually with a tripod and "driver", with pipe sections added as needed. A driver is a weighted pipe that slides over the pipe being driven and is repeatedly dropped on it. When groundwater is encountered, the well is washed of sediment and a pump installed.

Drilled wells

Drilled wells can be excavated by simple hand drilling methods (augering, sludging, jetting, driving, hand percussion) or machine drilling (rotary, percussion, down the hole hammer). Drilled wells can get water from a much deeper level than can dug wells - often up to several hundred metres.

Drilled wells with electric pumps are currently used throughout the world, typically in rural or sparsely populated areas, though many urban areas are supplied partly by municipal wells.

Drilled wells are typically created using either top-head rotary style, table rotary, or cable tool drilling machines, all of which use drilling stems that are turned to create a cutting action in the formation, hence the term 'drilling'. Most shallow well drilling machines are mounted on large trucks, trailers, or tracked vehicle carriages. Water wells typically range from 10 to 600 feet (180 m), but in some areas can go deeper than 3,000 feet (910 m).

Rotary drilling machines use a segmented steel drilling string, typically made up of 20-foot (6.1 m) sections of galvanized steel tubing that are threaded together, with a bit or other drilling device at the bottom end. Some rotary drilling machines are designed to install (by driving or drilling) a steel casing into the well in conjunction with the drilling of the actual bore hole. Air and/or water is used as a circulation fluid to displace cuttings and cool bits during the drilling.

Another form of rotary style drilling, termed 'mud rotary', makes use of a specially made mud, or drilling fluid, which is constantly being altered during the drill so that it can consistently create enough hydraulic pressure to hold the side walls of the bore hole open, regardless of the presence of a casing in the well. Typically, boreholes drilled into solid rock are not cased until after the drilling process is completed, regardless of the machinery used.

The oldest form of drilling machinery is the Cable Tool, still used today. Specifically designed to raise and lower a bit into the bore hole, the 'spudding' of the drill causes the bit to be raised and dropped onto the bottom of the hole, and the design of the cable causes the bit to twist at approximately 1/4 revolution per drop, thereby creating a drilling action. Unlike rotary drilling, cable tool drilling requires the drilling action to be stopped so that the bore hole can be bailed or emptied of drilled cuttings.

Drilled wells are usually cased with a factory-made pipe, typically steel (in air rotary or cable tool drilling) or plastic/PVC (in mud rotary wells, also present in wells drilled into solid rock). The casing is constructed by welding, either chemically or thermodynamically, segments of casing together. If the casing is installed during the drilling, most drills will drive the casing into the ground as the bore hole advances, while some newer machines will actually allow for the casing to be rotated and drilled into the formation in a similar manner as the bit advancing just below. PVC or plastic is typically welded and then lowered into the drilled well, vertically stacked with their ends nested and either glued or splined together. The sections of casing are usually 20' (6 m) or more in length, and 6"-12" (15 to 30 cm) in diameter, depending on the intended use of the well and local groundwater conditions.

Surface contamination of wells in the United States is typically controlled by the use of a 'surface seal'. A large hole is drilled to a predetermined depth or to a confining formation (clay or bedrock, for example), and then a smaller hole for the well is completed from that point forward. The well is typically cased from the surface down into the smaller hole with a casing that is the same diameter as that hole. The annular space between the large bore hole and the smaller casing is filled with bentonite clay, concrete, or other sealant material.

This creates an impermeable seal from the surface to the next confining layer that keeps contaminants from traveling down the outer sidewalls of the casing or borehole and into the aquifer. In addition, wells are typically capped with either an engineered well cap or seal that vents air through a screen into the well, but keeps insects, small animals, and unauthorized persons from accessing the well.

At the bottom of wells, based on formation, a screening device, filter pack, slotted casing, or open bore hole is left to allow the flow of water into the well. Constructed screens are typically used in unconsolidated formations (sands, gravels, etc.), allowing water and a percentage of the formation to pass through the screen. Allowing some material to pass through creates a large area filter out of the rest of the formation, as the amount of material present to pass into the well slowly decreases and is removed from the well.

Rock wells are typically cased with a PVC liner/casing and screen or slotted casing at the bottom, this is mostly present just to keep rocks from entering the pump assembly. Some wells utilize a 'filter pack' method, where an undersized screen or slotted casing is placed inside the well and a filter medium is packed around the screen, between the screen and the borehole or casing. This allows the water to be filtered of unwanted materials before entering the well and pumping zone.

Classification

There are two broad classes of drilled-well types, based on the type of aquifer the well is in:

- Shallow or unconfined wells are completed in the uppermost saturated aquifer at that location (the upper unconfined aquifer).
- Deep or confined wells are sunk through an impermeable stratum into an aquifer that is sandwiched between two impermeable strata (aquitards or aquicludes). The majority of deep aquifers are classified as artesian because the hydraulic head in a confined well is higher than the level of the top of the aquifer. If the hydraulic head in a confined well is higher than the land surface it is a "flowing" artesian well (named after Artois in France).

Two additional broad classes of well types may be distinguished, based on the use of the well:

- production or pumping wells, are large diameter (> 15 cm in diameter) cased (metal, plastic, or concrete) water wells, constructed for extracting water from the aquifer by a pump (if the well is not artesian).
- monitoring wells or piezometers, are often smaller diameter wells used to monitor the hydraulic head or sample the groundwater for chemical constituents. Piezometers are monitoring wells completed over a very short section of aquifer. Monitoring wells can also be completed at multiple levels, allowing discrete samples or measurements to be made at different vertical elevations at the same map location.

Obviously, a well constructed for pumping groundwater can be used passively as a monitoring well and a small diameter well can be pumped, but this distinction by use is common.

Well Siting

Before excavation, information about the geology, water table depth, seasonal fluctuations, recharge area and rate must be found. This work is typically done by a hydrogeologist, or a groundwater surveyor using a variety of tools including Electro-Seismic Surveying any available information from nearby wells, geologic maps sometimes geophysical imaging.

Well Contamination

Shallow pumping wells can often supply drinking water at a very low cost, but because impurities from the surface easily reach shallow sources, a greater risk of contamination occurs for these wells when they are compared to deeper wells. Dug and driven wells are relatively easy to contaminate, and dug wells are unreliable in most of the U.S. The quality of the well water can be significantly increased by lining the well, sealing the well head, fitting a self-priming hand pump, constructing an apron, ensuring the area is kept clean and free from stagnant water and animals, moving sources of contamination (latrines, garbage pits) and carrying out hygiene education. It is important that the well is cleaned with 1% chlorine solution after construction and periodically every 6 months.

Microorganisms

Most of the bacteria, viruses, parasites, and fungi that contaminate well water comes from fecal material from humans and other animals. Common bacterial contaminants include *E. coli*, Salmonella, Shigella, and Campylobacter jejuni. Common viral contaminants include norovirus, sapovirus, rotavirus, enteroviruses, and hepatitis A and E. Parasites include Giardia lamblia, Cryptosporidium, Cyclospora cayetanensis, and microsporidia.

Chemicals

Chemical contamination is a common problem with groundwater. Nitrates from sewage or fertilizer are a particular problem for children. Pesticides and volatile organic compounds from gasoline, dry-cleaning, and many other sources are the most commonly occurring pollutant chemicals in the U.S., and may be identifiable in more than a third of all U.S. wells, although this is mostly at levels below U.S. water standards. Other notable chemical contaminants include the fuel additive methyl tert-butyl ether (MTBE), and perchlorate from rocket fuel, airbag inflators, and other artificial and natural sources.

Several minerals are also contaminants, including lead leached from brass fittings or old lead pipes, chromium VI from electroplating and other sources, naturally occurring arsenic, radon, and uranium-all of which can cause cancer-and naturally occurring fluoride, which is desirable in low quantities to prevent tooth decay, but can cause dental fluorosis in higher concentrations.

Some chemicals are commonly present in water wells at levels that are not toxic, but can cause other problems. Calcium and magnesium cause what is known as hard water, which can precipitate and clog pipes or burn out water heaters. Iron and manganese can appear as dark flecks that stain clothing and plumbing, and can promote the growth of iron and manganese bacteria that can form slimy black colonies that clog pipes.

Mitigation

Cleanup of contaminated groundwater tends to be very costly. Effective remediation of groundwater is generally very difficult.

Contamination of groundwater from surface and subsurface sources can usually be dramatically reduced by correctly centering the casing during construction and filling the casing annulus with an appropriate sealing material.

The sealing material (grout) should be placed from immediately above the production zone back to surface, because, in the absence of a correctly constructed casing seal, contaminated fluid can travel into the well through the casing annulus. Centering devices are important (usually 1 per length of casing or at maximum intervals of 9 m) to ensure that the grouted annular space is of even thickness.

Upon the construction of a new test well, it is considered best practice to invest in a complete battery of chemical and biological tests on the well water in question. Point-of-use treatment is available for individual properties and treatment plants are often constructed for municipal water supplies that suffer from contamination. Most of these treatment methods involve the filtration of the contaminants of concern, and additional protection may be garnered by installing well-casing screens only at depths where contamination is not present.

Well water for personal use is often filtered with reverse osmosis water processors; this process can remove very small particles. A simple, effective way of killing microorganisms is to bring the water to a full boil for one to three minutes, depending on location. A household well contaminated by microorganisms can initially be treated by shock chlorination using bleach, generating concentrations hundreds of times greater than found in community water systems; however, this will not fix any structural problems that led to the contamination and generally requires some expertise and testing for effective application.

Environmental problems

A risk with the placement of water wells is soil salination. This problem occurs when the watertable of the soil begins to drop and salt begins to accumulate as the soil begins to dry out. Another environmental problem that is very prevalent in water well drilling is the potential for methane to seep through.

Soil Salination

The potential for soil salination is a large risk when choosing the placement of water wells. Soil Salination is caused when the water table of the soil drops over time and salt begins to accumulate. In turn, the increased amount of salt begins to dry the soil out. This is a very detrimental problem because the increased level of salt in the soil can result in the degradation of soil and can be very harmful to vegetation.

Methane

Methane, an asphyxiate, is a chemical compound that is the main component of natural gas. When methane mixes with oxygen, it has the ability to reduce oxygen to harmfully low levels in a small space or to form an explosion. This explosive quality is what poses such a danger in regards to the drilling and placement of water wells.

Low levels of methane in drinking water are not considered toxic. When methane seeps into a water supply, it is commonly referred to as "methane migration." This can be caused by old natural gas wells near water well systems becoming abandoned and no longer monitored.

ANCIENT WELL TECHNOLOGIES

The earliest wells are known from the Neolithic. In the submerged Pre-Pottery Neolithic B settlement of Atlit Yam in Israel, dated to 8100-7500 BC, a well has been found, which so far is the oldest known. Other PPNB wells (7-8 m deep) are known from Kissonerga-Mylouthkia on Cyprus and maybe shallower examples from Shillourokambos as well.

Wood-lined wells are known from the early Neolithic Linear Pottery culture, for example in Kückhoven, dated 5090 BC and Eythra, dated 5200 BC in Germany and Schletz in Austria. The early Mesolithic site of Friesack in Germany has yielded a shallow pit with the remains of a birch-bark container that may have been a shallow artificial well.

Australian Aborigines relied on wells to survive the harsh Australian desert. They would dig down, scooping out sand and mud to reach clean water, then cover the source with spinifex to prevent spoilage. Non-aborigines call these native wells, soaks or soakages.

Stepwells are common in the west of India. In these wells, the water may be reached by descending a set of steps. They may be covered and are often of architectural significance. Many stepwells were also used for leisure, providing relief from the daytime heat.

A qanat

A qanat is an ancient water collection system made up of a series of wells and linked underground water channels that collects flowing water from a source usually a distance away, stores it, and then brings the water to the surface using gravity. Much of the population of Iran and other arid countries in Asia and North Africa historically depended upon the water from qanats; the areas of population corresponded closely to the areas where qanats are possible.

In Egypt, shadoofs and sakiehs are used. When compared to each other however, the Sakkieh is much more efficient, as it can bring up water from a depth of 10 meters (versus the 3 meters of the shadoof). The Sakieh is the Egyptian version of the Noria. From the Iron Age onwards, wells are common archaeological features, both with wooden shafts and shaft linings made from wickerwork. Lately, however, the described wells/pumps are no longer very efficient and can be replaced by either handpumps or treadle pumps. Another alternative is the use of self-dug wells, electrical deep-well pumps (for higher depths). Appropriate technology organizations as Practical Action are now supplying information on how to build/set-up (diy) handpumps and treadle pumps in practice.

Flood

A flood is an overflow of an expanse of water that submerges land. The EU Floods directive defines a flood as a temporary covering by water of land not normally covered by water. In the sense of "flowing water", the word may also be applied to the inflow of the tide. Flooding may result from the volume of water within a body of water, such as a river or lake, which overflows or breaks levees, with the result that some of the water escapes its usual boundaries.

While the size of a lake or other body of water will vary with seasonal changes in precipitation and snow melt, it is not a significant flood unless such escapes of water endanger land areas used by man like a village, city or other inhabited area.

Floods can also occur in rivers, when flow exceeds the capacity of the river channel, particularly at bends or meanders. Floods often cause damage to homes and businesses if they are placed in natural flood plains of rivers. While flood damage can be virtually eliminated by moving away from rivers and other bodies of water, since time out of mind, people have lived and worked by the water to seek sustenance and capitalize on the gains of cheap and easy travel and commerce by being near water. That humans continue to inhabit areas threatened by flood damage is evidence that the perceived value of living near the water exceeds the cost of repeated periodic flooding.

Dozens of villages were inundated when rain pushed the rivers of northwestern Bangladesh over their banks in early October 2005. The Moderate Resolution Imaging Spectroradiometer (MODIS) on NASA's Terra satellite captured the top image of the flooded Ghaghat and Atrai Rivers on October 12, 2005. The deep blue of the rivers is spread across the countryside in the flood image.

Riverine

- Slow kinds: Runoff from sustained rainfall or rapid snow melt exceeding the capacity of a river's channel. Causes include heavy rains from monsoons, hurricanes and tropical depressions, foreign winds and warm rain affecting snow pack. Unexpected drainage obstructions such as landslides, ice, or debris can cause slow flooding upstream of the obstruction.
- Fast kinds: include flash floods resulting from convective precipitation (intense thunderstorms) or sudden release from an upstream impoundment created behind a dam, landslide, or glacier.

Estuarinet

- Commonly caused by a combination of sea tidal surges caused by storm-force winds. A storm surge, from either a tropical cyclone or an extratropical cyclone, falls within this category.

Coastal

- Caused by severe sea storms, or as a result of another hazard (e.g. tsunami or hurricane). A storm surge, from either a tropical cyclone or an extratropical cyclone, falls within this category.

Catastrophic

- Caused by a significant and unexpected event e.g. dam breakage, or as a result of another hazard (e.g. earthquake or volcanic eruption).

Human-induced

Accidental damage by workmen to tunnels or pipes.

Muddy

- A muddy flood is generated by runoff on crop land.

A muddy flood is produced by an accumulation of runoff generated on cropland. Sediments are then detached by runoff and carried as suspended matter or bed load. Muddy runoff is more likely detected when it reaches inhabited areas.

Muddy floods are therefore a hill slope process, and confusion with mudflows produced by mass movements should be avoided.

Other

- Floods can occur if water accumulates across an impermeable surface (e.g. from rainfall) and cannot rapidly dissipate (i.e. gentle orientation or low evaporation).
- A series of storms moving over the same area.
- Dam-building beavers can flood low-lying urban and rural areas, often causing significant damage.

Effects

Primary effects

- Physical damage - Can damage any type of structure, including bridges, cars, buildings, sewerage systems, roadways, and canals.

Secondary effects

- Water supplies - Contamination of water. Clean drinking water becomes scarce.
- Diseases - Unhygienic conditions. Spread of water-borne diseases.
- Crops and food supplies - Shortage of food crops can be caused due to loss of entire harvest. However, lowlands near rivers depend upon river silt deposited by floods in order to add nutrients to the local soil.
- Trees - Non-tolerant species can die from suffocation.

Control

In many countries across the world, rivers prone to floods are often carefully managed. Defenses such as levees, bunds, reservoirs, and weirs are used to prevent rivers from bursting their banks. When these defenses fail, emergency measures such as sandbags or portable inflatable tubes are used. Coastal flooding has been addressed in Europe and the Americas with coastal defences, such as sea walls, beach nourishment, and barrier islands.

DROUGHT

A drought (or drouth [archaic]) is an extended period of months or years when a region notes a deficiency in its water supply. Generally, this occurs when a region receives consistently below average precipitation. It can have a substantial impact on the ecosystem and agriculture of the affected region. Although droughts can persist for several years, even a short, intense drought can cause significant damage and harm the local economy.

This global phenomenon has a widespread impact on agriculture. The United Nations estimates that an area of fertile soil the size of Ukraine is lost every year because of drought, deforestation, and climate instability. Lengthy periods of drought have long been a key trigger for mass migration and played a key role in a number of ongoing migrations and other humanitarian crises in the Horn of Africa and the Sahel.

Consequences

Periods of drought can have significant environmental, agricultural, health, economic and social consequences. The effect varies according to vulnerability. For example, subsistence farmers are more likely to migrate during drought because they do not have alternative food sources. Areas with populations that depend on subsistence farming as a major food source are more vulnerable to drought-triggered famine.

Drought can also reduce water quality, because lower water flows reduce dilution of pollutants and increase contamination of remaining water sources. Common consequences of drought include:

- Diminished crop growth or yield productions and carrying capacity for livestock
- Dust bowls, themselves a sign of erosion, which further erode the landscape
- Dust storms, when drought hits an area suffering from desertification and erosion
- Famine due to lack of water for irrigation
- Habitat damage, affecting both terrestrial and aquatic wildlife
- Malnutrition, dehydration and related diseases
- Mass migration, resulting in internal displacement and international refugees

- Reduced electricity production due to insufficient available coolant for power stations, and reduced water flow through hydroelectric dams
- Shortages of water for industrial users
- Snakes migration and increases in snakebites
- Social unrest
- War over natural resources, including water and food
- Wildfires, such as Australian bushfires, are more common during times of drought

Globally

Drought is a normal, recurring feature of the climate in most parts of the world. It is among the earliest documented climatic events, present in the Epic of Gilgamesh and tied to the biblical story of Joseph's arrival in and the later Exodus from Ancient Egypt. Hunter-gatherer migrations in 9,500 BC Chile have been linked to the phenomenon, as has the exodus of early man out of Africa and into the rest of the world around 135,000 years ago.

Modern peoples can effectively mitigate much of the impact of drought through irrigation and crop rotation. Failure to develop adequate drought mitigation strategies carries a grave human cost in the modern era, exacerbated by ever-increasing population densities.

Regions

Recurring droughts leading to desertification in the Horn of Africa have created grave ecological catastrophes, prompting massive food shortages, still recurring. To the north-west of the Horn, the Darfur conflict in neighboring Sudan, also affecting Chad, was fueled by decades of drought; combination of drought, desertification and overpopulation are among the causes of the Darfur conflict, because the Arab Baggara nomads searching for water have to take their livestock further south, to land mainly occupied by non-Arab farming peoples.

Approximately 2.4 billion people live in the drainage basin of the Himalayan rivers. India, China, Pakistan, Bangladesh, Nepal and Myanmar could experience floods followed by droughts in coming decades. Drought in India affecting the Ganges is of particular concern, as it provides drinking water and agricultural irrigation for more than 500 million people. The west coast of North America, which gets much of its water from glaciers in mountain ranges such as the Rocky Mountains and Sierra Nevada, also would be affected.

In 2005, parts of the Amazon basin experienced the worst drought in 100 years. A 23 July 2006 article reported Woods Hole Research Center results showing that the forest in its present form could survive only three years of drought. Scientists at the Brazilian National Institute of Amazonian Research argue in the article that this drought response, coupled with the effects of deforestation on regional climate, are pushing the rainforest towards a "tipping point" where it would irreversibly start to die. It concludes that the rainforest is on the brink of being turned into savanna or desert, with catastrophic consequences for the world's climate. According to the WWF, the combination of climate change and deforestation increases the drying effect of dead trees that fuels forest fires.

By far the largest part of Australia is desert or semi-arid lands commonly known as the outback. A 2005 study by Australian and American researchers investigated the desertification of the interior, and suggested that one explanation was related to human settlers who arrived about 50,000 years ago. Regular burning by these settlers could have prevented monsoons from reaching interior Australia. In June 2008 it became known that an expert panel had warned of long term, maybe irreversible, severe ecological damage for the whole Murray-Darling basin if it does not receive sufficient water by October. Australia could experience more severe droughts and they could become more frequent in the future, a government-commissioned report said on July 6, 2008. The Australian of the year 2007, environmentalist Tim Flannery, predicted that unless it made drastic changes, Perth in Western Australia could become the world's first ghost metropolis, an abandoned city with no more water to sustain its population.

East Africa currently faces its worst drought in decades, with crops and livestock destroyed. The U.N. World Food Programme recently said that nearly four million Kenyans urgently needed food.

Causes

Generally, rainfall is related to the amount of water vapor in the atmosphere, combined with the upward forcing of the air mass containing that water vapor. If either of these are reduced, the result is a drought. This can be triggered by an above average prevalence of high pressure systems, winds carrying continental, rather than oceanic air masses (i.e. reduced water content), and ridges of high pressure areas form with behaviors which prevent or restrict the developing of thunderstorm activity or rainfall over one certain region. Oceanic and atmospheric weather cycles such as the El Niño-Southern Oscillation (ENSO) make drought a regular recurring feature of the Americas along the Pacific coast and Australia. Guns, Germs, and Steel author Jared Diamond sees the stark impact of the multi-year ENSO cycles on Australian weather patterns as a key reason that Australian aborigines remained a hunter-gatherer society rather than adopting agriculture. Another climate oscillation known as the North Atlantic Oscillation has been tied to droughts in northeast Spain.

Human activity can directly trigger exacerbating factors such as over farming, excessive irrigation, deforestation, and erosion adversely impact the ability of the land to capture and hold water. While these tend to be relatively isolated in their scope, activities resulting in global climate change are expected to trigger droughts with a substantial impact on agriculture throughout the world, and especially in developing nations. Overall, global warming will result in increased world rainfall. Along with drought in some areas, flooding and erosion will increase in others. Paradoxically, some proposed solutions to global warming that focus on more active techniques, solar radiation management through the use of a space sunshade for one, may also carry with them increased chances of drought.

Types of drought

As a drought persists, the conditions surrounding it gradually worsen and its impact on the local population gradually increases. People tend to define droughts in three main ways:

1. Meteorological drought is brought about when there is a prolonged period with less than average precipitation. Meteorological drought usually precedes the other kinds of drought.
2. Agricultural droughts are droughts that affect crop production or the ecology of the range. This condition can also arise independently from any change in precipitation levels when soil conditions and erosion triggered by poorly planned agricultural endeavors cause a shortfall in water available to the crops. However, in a traditional drought, it is caused by an extended period of below average precipitation.
3. Hydrological drought is brought about when the water reserves available in sources such as aquifers, lakes are reservoirs fall below the statistical average. Hydrological drought tends to show up more slowly because it involves stored water that is used but not replenished. Like an agricultural drought, this can be triggered by more than just a loss of rainfall. For instance, Kazakhstan was recently awarded a large amount of money by the World Bank to restore water that had been diverted to other nations from the Aral Sea under Soviet rule. Similar circumstances also place their largest lake, Balkhash, at risk of completely drying out.

Drought protection and relief

The effects of the drought brought on by El Niño. Waiting for water distribution (Ebeye, Marshall Islands.)

Strategies for drought protection, mitigation or relief include:

- Dams - many dams and their associated reservoirs supply additional water in times of drought.
- Cloud seeding - an artificial technique are induce rainfall.
- Desalination of sea water for irrigation or consumption.
- Drought monitoring - Continuous observation of rainfall levels and comparisons with current usage levels can help prevent man-made drought. For instance, analysis of water usage in Yemen has revealed that their water table (underground water level) is put at grave risk by over-use to fertilize their Khat crop. Careful monitoring of moisture levels can also help predict increased risk for wildfires, using such metrics as the Keetch-Byram Drought Index or Palmer Drought Index.
- Land use - Carefully planned crop rotation can help to minimize erosion and allow farmers to plant less water-dependent crops in drier years.
- Outdoor water-use restriction - Regulating the use of sprinklers, hoses or buckets on outdoor plants, filling pools, and other water-intensive home maintenance tasks.
- Rainwater harvesting - Collection and storage of rainwater from roofs or other suitable catchments.
- Recycled water - Former wastewater (sewage) that has been treated and purified for reuse.
- Transvasement - Building canals or redirecting rivers as massive attempts at irrigation in drought-prone areas.

Surface runoff

Surface runoff is the water flow that occurs when soil is infiltrated to full capacity and excess water from rain, meltwater, or other sources flows over the land. This is a major component of the water cycle. Runoff that occurs on surfaces before reaching a channel is also called a nonpoint source. If a nonpoint source contains man-made contaminants, the runoff is called nonpoint source pollution. A land area which produces runoff that drains to a common point is called a watershed. When runoff flows along the ground, it can pick up soil contaminants including, but not limited to petroleum, pesticides, or fertilizers that become discharge or nonpoint source pollution.

Surface runoff can be generated either by rainfall or by the melting of snow, or glaciers. Snow and glacier melt occur only in areas cold enough for these to form permanently. Typically snowmelt will peak in the spring and glacier melt in the summer, leading to pronounced flow maxima in rivers affected by them. The determining factor of the rate of melting of snow or glaciers is both air temperature and the duration of sunlight. In high mountain regions, streams frequently rise on sunny days and fall on cloudy ones for this reason.

In areas where there is no snow, runoff will come from rainfall. However, not all rainfall will produce runoff because storage from soils can absorb light showers. On the extremely ancient soils of Australia and Southern Africa, proteoid roots with their extremely dense networks of root hairs can absorb so much rainwater as to prevent runoff even when substantial amounts of rain fall. In these regions, even on less infertile cracking clay soils, high amounts of rainfall and low potential evaporation are needed to generate any surface runoff, leading to specialised adaptations to extremely variable (usually ephemeral) streams.

Infiltration excess overland flow

This occurs when the rate of rainfall on a surface exceeds the rate at which water can infiltrate the ground, and any depression storage has already been filled. This is called infiltration excess overland flow, Hortonian overland flow (after Robert E. Horton), or unsaturated overland flow. This more commonly occurs in arid and semi-arid regions, where rainfall intensities are high and the soil infiltration capacity is reduced because of surface sealing, or in paved areas. This occurs largely in city areas where pavements prevent water infiltration.

Saturation excess overland flow

When the soil is saturated and the depression storage filled, and rain continues to fall, the rainfall will immediately produce surface runoff. The level of antecedent soil moisture is one factor affecting the time until soil becomes saturated. This runoff is called saturation excess overland flow or saturated overland flow.

Antecedent soil moisture

Soil retains a degree of moisture after a rainfall. This residual water moisture affects the soil's infiltration capacity. During the next rainfall event, the infiltration capacity will cause the soil to be saturated at a different rate. The higher the level of antecedent soil moisture, the more quickly the soil becomes saturated. Once the soil is saturated, runoff occurs.

Subsurface return flow

After water infiltrates the soil on an up-slope portion of a hill, the water may flow laterally through the soil, and exfiltrate (flow out of the soil) closer to a channel. This is called subsurface return flow or throughflow.

As it flows, the amount of runoff may be reduced in a number of possible ways: a small portion of it may evapotranspire; water may become temporarily stored in microtopographic depressions; and a portion of it may become run-on, which is the infiltration of runoff as it flows overland. Any remaining surface water eventually flows into a receiving water body such as a river, lake, estuary or ocean.

Human impact on surface runoff

Urbanization increases surface runoff, by creating more impervious surfaces such as pavement and buildings, that do not allow percolation of the water down through the soil to the aquifer. It is instead forced directly into streams or storm water runoff drains, where erosion and siltation can be major problems, even when flooding is not. Increased runoff reduces groundwater recharge, thus lowering the water table and making droughts worse, especially for farmers and others who depend on the water wells.

When anthropogenic contaminants are dissolved or suspended in runoff, the human impact is expanded to create water pollution. This pollutant load can reach various receiving waters such as streams, rivers, lakes, estuaries and oceans with resultant water chemistry changes to these water systems and their related ecosystems. A 2008 report by the United States National Research Council identified urban stormwater as a leading source of water quality problems in the U.S.

Effects of surface runoff

Erosion & deposition

Surface runoff causes erosion of the Earth's surface; deposition is the depositing of erosion. There are four principal types of erosion: splash erosion, gully erosion, sheet erosion and stream bed erosion. Splash erosion is the result of mechanical collision of raindrops with the soil surface. Dislodged soil particles becoming suspended in the surface runoff and carried into streams and rivers. Gully erosion occurs when the power of runoff is strong enough that it cuts a well defined channel. These channels can be as small as one centimeter wide or as large as several meters.

Sheet erosion is the overland transport of runoff without a well defined channel. In the case of gully erosion, large amounts of material can be transported in a small time period. Stream bed erosion is the attrition of stream banks or bottoms by rapidly flowing rivers or creeks.

Reduced crop productivity usually results from erosion, and these effects are studied in the field of soil conservation. The soil particles carried in runoff vary in size from about .001 millimeter to 1.0 millimeter in diameter. Larger particles settle over short transport distances, whereas small particles can be carried over long distances suspended in the water column. Erosion of silty soils that contain smaller particles generates turbidity and diminishes light transmission, which disrupts aquatic ecosystems.

Entire sections of countries have been rendered unproductive by erosion. On the high central plateau of Madagascar, approximately ten percent of that country's land area, virtually the entire landscape is devoid of vegetation, with erosive gully furrows typically in excess of 50 meters deep and one kilometer wide.

Shifting cultivation is a farming system which sometimes incorporates the slash and burn method in some regions of the world. Erosion cause loss of the fertile top soil and reduces the its fertility and quality of the agricultural produce. Modern industrial farming is another major cause of erosion. In some areas in the American corn belt, more than 50 percent of the original topsoil has been carried away within the last 100 years.

ENVIRONMENTAL IMPACTS

The principal environmental issues associated with runoff are the impacts to surface water, groundwater and soil through transport of water pollutants to these systems. Ultimately these consequences translate into human health risk, ecosystem disturbance and aesthetic impact to water resources. Some of the contaminants that create the greatest impact to surface waters arising from runoff are petroleum substances, herbicides and fertilizers.

Quantitative uptake by surface runoff of pesticides and other contaminants has been studied since the 1960s, and early on contact of pesticides with water was known to enhance phytotoxicity. In the case of surface waters, the impacts translate to water pollution, since the streams and rivers have received runoff carrying various chemicals or sediments. When surface waters are used as potable water supplies, they can be compromised regarding health risks and drinking water aesthetics (that is, odor, color and turbidity effects).

Contaminated surface waters risk altering the metabolic processes of the aquatic species that they host; these alterations can lead to death, such as fish kills, or alter the balance of populations present. Other specific impacts are on animal mating, spawning, egg and larvae viability, juvenile survival and plant productivity. Some researches show surface runoff of pesticides, such as DDT, can alter the gender of fish species genetically, which transforms male into female fish.

In the case of groundwater, the main issue is contamination of drinking water, if the aquifer is abstracted for human use. Regarding soil contamination, runoff waters can have two important pathways of concern. Firstly, runoff water can extract soil contaminants and carry them in the form of water pollution to even more sensitive aquatic habitats. Secondly, runoff can deposit contaminants on pristine soils, creating health or ecological consequences.

Flooding

Flooding occurs when a watercourse is unable to convey the quantity of runoff flowing downstream. The frequency with which this occurs is described by a return period. Flooding is a natural process, which maintains ecosystem composition and processes, but it can also be altered by land use changes such as river engineering. Floods can be both beneficial to societies or cause damage. Agriculture along the Nile floodplain took advantage of the seasonal flooding that deposited nutrients beneficial for crops.

However, as the number and susceptibility of settlements increase, flooding increasingly becomes a natural hazard. Adverse impacts span loss of life, property damage, contamination of water supplies, loss of crops, and social dislocation and temporary homelessness. Floods are among the most devastating of natural disasters.

Agricultural issues

A common context of run-off deals with agriculture. When farmland is tilled and bare soil is revealed, rainwater carries billions of tons of topsoil into waterways each year, causing loss of valuable topsoil and adding sediment to produce turbidity in surface waters.

The other context of agricultural issues involves the transport of agricultural chemicals (nitrates, phosphates, pesticides, herbicides etc) via surface runoff. This result occurs when chemical use is excessive or poorly timed with respect to high precipitation. The resulting contaminated runoff represents not only a waste of agricultural chemicals, but also an environmental threat to downstream ecosystems. The alternative to conventional farming is organic farming which eliminates chemical usage.

Measurement and mathematical modeling

Runoff is analyzed by using mathematical models in combination with various water quality sampling methods. Measurements can be made using continuous automated water quality analysis instruments targeted on pollutants such as specific organic or inorganic chemicals, pH, turbidity etc. or targeted on secondary indicators such as dissolved oxygen. Measurements can also be made in batch form by extracting a single water sample and conducting any number of chemical or physical tests on that sample.

In the 1950s or earlier hydrology transport models appeared to calculate quantities of runoff, primarily for flood forecasting. Beginning in the early 1970s computer models were developed to analyze the transport of runoff carrying water pollutants, which considered dissolution rates of various chemicals, infiltration into soils and ultimate pollutant load delivered to receiving waters.

One of the earliest models addressing chemical dissolution in runoff and resulting transport was developed in the early 1970s under contract to the United States Environmental Protection Agency (EPA). This computer model formed the basis of much of the mitigation study that led to strategies for land use and chemical handling controls.

Other computer models have been developed (such as the DSSAM Model) that allow surface runoff to be tracked through a river course as reactive water pollutants. In this case the surface runoff may be considered to be a line source of water pollution to the receiving waters.

Mitigation and treatment

Mitigation of adverse impacts of runoff can take several forms:

- Land use development controls aimed at minimizing impervious surfaces in urban areas
- Erosion controls for farms and construction sites
- Flood control programs
- Chemical use and handling controls in agriculture, landscape maintenance, industrial use, etc.

Land use controls. Many world regulatory agencies have encouraged research on methods of minimizing total surface runoff by avoiding unnecessary hardscape.

Many municipalities have produced guidelines and codes (zoning and related ordinances) for land developers that encourage minimum width sidewalks, use of pavers set in earth for driveways and walkways and other design techniques to allow maximum water infiltration in urban settings. An example land use control program can be viewed at seen in the city of Santa Monica, California.

Erosion controls have appeared since medieval times when farmers realized the importance of contour farming to protect soil resources. Beginning in the 1950s these agricultural methods became increasingly more sophisticated. In the 1960s some state and local governments began to focus their efforts on mitigation of construction runoff by requiring builders to implement erosion and sediment controls (ESCs).

This included such techniques as: use of straw bales and barriers to slow runoff on slopes, installation of silt fences, programming construction for months that have less rainfall and minimizing extent and duration of exposed graded areas. Montgomery County, Maryland implemented the first local government sediment control program in 1965, and this was followed by a statewide program in Maryland in 1970.

Flood control programs as early as the first half of the twentieth century became quantitative in predicting peak flows of riverine systems. Progressively strategies have been developed to minimize peak flows and also to reduce channel velocities. Some of the techniques commonly applied are: provision of holding ponds (also called detention basins) to buffer riverine peak flows, use of energy dissipators in channels to reduce stream velocity and land use controls to minimize runoff.

Chemical use and handling. Following enactment of the U.S. Resource Conservation and Recovery Act (RCRA) in 1976, and later the Water Quality Act of 1987, states and cities have become more vigilant in controlling the containment and storage of toxic chemicals, thus preventing releases and leakage.

Methods commonly applied are: requirements for double containment of underground storage tanks, registration of hazardous materials usage, reduction in numbers of allowed pesticides and more stringent regulation of fertilizers and herbicides in landscape maintenance. In many industrial cases, pretreatment of wastes is required, to minimize escape of pollutants into sanitary or stormwater sewers.

The U.S. Clean Water Act (CWA) requires that local governments in urbanized areas (as defined by the Census Bureau) obtain stormwater discharge permits for their drainage systems. Essentially this means that the locality must operate a stormwater management program for all surface runoff that enters the municipal separate storm sewer system ("MS4"). EPA and state regulations and related publications outline six basic components that each local program must contain:

- Public education (informing individuals, households, businesses about ways to avoid stormwater pollution)
- Public involvement (support public participation in implementation of local programs)
- Illicit discharge detection & elimination (removing sanitary sewer or other non-stormwater connections to the MS4)
- Construction site runoff controls (i.e. erosion & sediment controls)
- Post-construction (i.e. permanent) stormwater management controls
- Pollution prevention and "good housekeeping" measures (e.g. system maintenance).

Other property owners which operate storm drain systems similar to municipalities, such as state highway systems, universities, military bases and prisons, are also subject to the MS4 permit requirements.

HYDROLOGICAL MODEL

A hydrologic model is a simplification of a real-world system (e.g., surface water, soil water, wetland, groundwater, estuary) that aids in understanding, predicting, and managing water resources. Both the flow and quality of water are commonly studied using hydrologic models.

Conceptual models

Conceptual models are commonly used to represent the important components (e.g., features, events, and processes) that relate hydrologic inputs to outputs. These components describe the important functions of the system of interest, and are often constructed using entities (stores of water) and relationships between these entitites (flows or fluxes between stores). The conceptual model is coupled with scenarios to describe specific events (either input or outcome scenarios).

For example, a watershed model could be represented using tributaries as boxes with arrows pointing toward a box that represents the main river. The conceptual model would then specify the important watershed features (e.g., land use, land cover, soils, subsoils, geology, wetlands, lakes), atmospheric exchanges (e.g., precipitation, evapotranspiration), human uses (e.g., agricultural, municipal, industrial, navigation, thermo- and hydro-electric power generation), flow processes (e.g., overland, interflow, baseflow, channel flow), transport processes (e.g., sediments, nutrients, pathogens), and events (e.g., low-, flood-, and mean-flow conditions).

Model scope and complexity is dependent on modeling objectives, with greater detail required if human or environmental systems are subject to greater risk. Systems modeling can be used for building conceptual models that are then populated using mathematical relationships.

Analog models

Prior to the advent of computer models, hydrologic modeling used analog models to simulate flow and transport systems. Unlike mathematical models that use equations to describe, predict, and manage hydrologic systems, analog models use non-mathematical approaches to simulate hydrology. Two general categories of analog models are common; scale analogs that use miniaturized versions of the physical system and process analogs that use comparable physics (e.g., electricity, heat, diffusion) to mimic the system of interest.

Scale analogs

Scale models offer a useful approximation of physical or chemical processes at a size that allows for greater ease of visualization. The model may be created in one (core, column), two (plan, profile), or three dimensions, and can be designed to represent a variety of specific initial and boundary conditions as needed to answer a question.

Scale models commonly use physical properties that are similar to their natural counterparts (e.g., gravity, temperature). Yet, maintaining some properties at their natural values can lead to erroneous predictions. Properties such as viscosity, friction, and surface area must be adjusted to maintain appropriate flow and transport behavior. This usually involves matching dimensionless ratios (e.g., Reynolds number, Froude number).

Groundwater flow can be visualized using a scale model built of acrylic and filled with sand, silt, and clay. Water and tracer dye may be pumped through this system to represent the flow of the simulated groundwater. Some physical aquifer models are between two and three dimensions, with simplified boundary conditions simulated using pumps and barriers.

Process analogs

Process analogs are used in hydrology to represent fluid flow using the similarity between Darcy's Law, Ohms Law, Fourier's Law, and Fick's Law. The analogs to fluid flow are the flux of electricity, heat, and solutes, respectively. The corresponding analogs to fluid potential are voltage, temperature, and solute concentration (or chemical potential). The analogs to hydraulic conductivity are electrical conductivity, thermal conductivity, and the solute diffusion coefficient.

An early process analog model was an electrical network model of an aquifer composed of resistors in a grid. Voltages were assigned along the outer boundary, and then measured within the domain. Electrical conductivity paper can also be used instead of resistors.

Statistical models

Statistical models are a type of mathematical model that are commonly used in hydrology to describe data, as well as relationships between data. Using statistical methods, hydrologists develop empirical relationships between observed variables, find trends in historical data, or forecast probable storm or drought events.

Moments

Statistical moments (e.g., mean, standard deviation, skewness, kurtosis) are used to describe the information content of data. These moments can then be used to determine an appropriate frequency distribution, which can then be used as a probability model. Two common techniques include L-moment ratios and Moment-Ratio Diagrams.

The frequency of extremal events, such as severe droughts and storms, often requires the use of distributions that focus on the tail of the distribution, rather than the data nearest the mean. These techniques, collectively known as extreme value analysis, provide a methodology for identifying the likelihood and uncertainty of extreme events. Examples of extreme value distributions include the Gumbel, Pearson, and Generalized Extreme Value. The standard method for determining peak discharge uses the log-Pearson Type III (log-gamma) distribution and observed annual flow peaks.

Correlation analysis

The degree and nature of correlation may be quantified, by using a method such as the Pearson correlation coefficient, autocorrelation, or the T-test. The degree of randomness or uncertainty in the model may also be estimated using stochastics, or residual analysis. These techniques may be used in the identification of flood dynamics, storm characterization, and groundwater flow in karst systems.

Regression analysis is used in hydrology to determine whether a relationship may exist between independent and dependent variables. Bivariate diagrams are the most commonly used statistical regression model in the physical sciences, but there are a variety of models available from simplistic to complex. In a bivariate diagram, a linear or higher-order model may be fitted to the data.

Factor Analysis and Principal Component Analysis are multivariate statistical procedures used to identify relationships between hydrologic variables,.

Convolution is a mathematical operation on two different functions to produce a third function. With respect to hydrologic modeling, convolution can be used to analyze stream discharge's relationship to precipitation. Convolution is used to predict discharge downstream after a precipitation event. This type of model would be considered a "lag convolution", because of the predicting of the "lag time" as water moves through the watershed using this method of modeling.

Time-series analysis is used to characterize temporal correlation within a data series as well as between different time series. Many hydrologic phenomena are studied within the context of historical probability. Within a temporal dataset, event frequencies, trends, and comparisons may be made by using the statistical techniques of time series analysis. The questions that are answered through these techniques are often important for municipal planning, civil engineering, and risk assessments. Markov Chains are a mathematical technique for determine the probability of a state or event based on a previous state or event. The event must be dependent, such as rainy weather. Markov Chains were first used to model rainfall event length in days in 1976, and continues to be used for flood risk assessment and dam management.

APPLICATIONS OF ENGINEERING HYDROLOGY

Hydrology is used to find out maximum probable flood at proposed sites e.g. Dams.

The variation of water production from catchments can be calculated and described by hydrology.

Engineering hydrology enables us to find out the relationship between a catchments's surface water and groundwater resources

The expected flood flows over a spillway, at a highway Culvert, or in an urban storm drainage system can be known by this very subject.

It helps us to know the required reservoir capacity to assure adequate water for irrigation or municipal water supply in droughts condition.

It tells us what hydrologic hardware (e.g. rain gauges, stream gauges etc) and software (computer models) are needed for real-time flood forecasting

Used in connection with design and operations of hydraulic structure

Used in prediction of flood over a spillway, at highway culvert or in urban storm drainage

Used to assess the reservoir capacity required to assure adequate water for irrigation or municipal water supply during drought

Hydrology is an indispensable tool in planning and building hydraulic structures.

Hydrology is used for city water supply design which is based on catchments area, amount of rainfall, dry period, storage capacity, runoff evaporation and transpiration.

Dam construction, reservoir capacity, spillway capacity, sizes of water supply pipelines and affect of afforest on water supply schemes, all are designed on basis of hydrological equations.

4

IRRIGATION IN INDIA

Irrigation in India includes a network of major and minor canals from Indian rivers, groundwater well based systems, tanks, and other rainwater harvesting projects for agricultural activities. Of these groundwater system is the largest. In 2013–14, only about 36.7% of total agricultural land in India was reliably irrigated, and remaining 2/3 cultivated land in India is dependent on monsoons. 65% of the irrigation in India is from groundwater. Currently about 51% of the agricultural area cultivating food grains is covered by irrigation. The rest of the area is dependent on rainfall which is most of the times unreliable and unpredictable.

Indian government launched a demand side water management plan costing INR6000 crore or USD854 million across 8,350 water stressed villages of 78 districts in seven states – Gujarat, Haryana, Karnataka, Madhya Pradesh, Maharashtra, Rajasthan and Uttar Pradesh – over five years from 2021–2022 to 2026–27, with the view to harvest rainwater, enhance water table, water recharge rate with village panchayat level water management plans. Most of the canal irrigation is in the canal network of Ganges-Yamuna basin mainly in the states of Punjab, Haryana, and Uttar Pradesh and somewhat in Rajasthan and Bihar, while small local canal networks also exist in the south in Tamil Nadu, Karnataka, and Kerala. The largest canal in India is Indira Gandhi Canal, which is about 650 km long. India has an ambitious river linking national project to enhance the coverage of canal-irrigated area, reduce floods and water shortage.

Irrigation in India helps improve food security, reduce dependence on monsoons, improve agricultural productivity and create rural job opportunities. Dams used for irrigation projects help produce electricity and transport facilities, as well as provide drinking water supplies to a growing population, control floods and prevent droughts.

This most ambitious, much talked about Interlinking of Rivers (ILR) project, estimated to cost approximately $112 billion USD (in 2002 price level), has attracted more debate than consensus in India. The ILR proposal which is backed by the three branches of the Indian political system - legislature, executive and judiciary, has been dismissed by civil society organizations as well as the traditional water managers. The recently signed Memorandum of Understanding (MoU) on a river link between most populous neighboring provinces - Uttar Pradesh and Madhya Pradesh - is evidence that the mammoth ILR project is being pursued but in a decentralised manner. Since its inception, the rationale of the ILR project has been under severe scrutiny not only in India but also in South Asia due to its magnitude.

Does India need a project of such magnitude to address its water management challenges? Is there any alternative to this grandiose intention in India? Grand water schemes are not new in the world. There are number of instances on this planet where human interventions using modern technology have redrawn otherwise natural river courses to address a water crisis.

The gigantic South-to-North water diversion project underway in China, large scale water diversions away from the Aral Sea in former Soviet Union, the Irtysh-Karaganda canal project in Kazakhstan, Israel's National Water Carrier, the South-Eastern Anatolia Project in Turkey (also known as GAP), Spain's National Hydrological Plan (now suspended) are the few examples of bold forays of water resources engineering. In similar fashion the Indian government has developed a proposal to tackle its peculiar climatic contradiction of drought and flood situations by proposing large scale interbasin water transfers and linking of rivers. The proposal has been presented as a major initiative, and as the definitive answer to India's future problems and needs.

HISTORY OF IRRIGATION IN INDIA

Ancient India

The earliest mentions of irrigation are found in Rigveda chapters 1.55, 1.85, 1.105, 7.9, 8.69 and 10.101. The Veda mentions only well-style irrigation, where kupa and avata wells once dug are stated to be always full of water, from which varatra (rope strap) and cakra (wheel) pull kosa (pails) of water. This water was, state the Vedas, led into surmi susira (broad channels) and from there into khanitrima (diverting channels) into fields.

Later, the 4th-century BCE Indian scholar Panini, mentions tapping several rivers for irrigation. The mentioned rivers include Sindhu, Suvastu, Varnu, Sarayu, Vipas and Chandrabhaga. Buddhist texts from the 3rd century BCE also mention irrigation of crops. Texts from the Maurya Empire era (3rd century BCE) mention that the state raised revenue from charging farmers for irrigation services from rivers.

Patanjali, in Yogasutra of about the 4th century CE, explains a technique of yoga by comparing it to "the way a farmer diverts a stream from an irrigation canal for irrigation". In Tamil Nadu, the Grand Anicut (canal) across the Kaveri river was implemented in the 3rd century CE, and the basic design is still used today.

Medieval era

The most widespread irrigation system in India was undertaken in India in the medieval period by the Sultanate rulers. Firoz Shah Tughlaq (1309-1388) built the most extensive canal irrigation system around the Indo-Gangetic doab and the region west of the river Yamuna in the fourteenth century. These canals provided vast resources of water to agricultural lands in northern India as well as vital supplies of water to urban and rural settlements. These irrigation projects were continued by the subsequent rulers of northern India, particularly the Mughal rulers till the early eighteenth century. The British built the colonial canal networks on these medieval canal systems.

Colonial era

In 1800, some 800,000 hectares were irrigated in India. The Britishers by 1940 built significant number of canals and irrigation systems in Uttar Pradesh, Bihar, Punjab, Assam and Orissa. The Ganges Canal reached 350 miles from Haridwar to Kanpur in Uttar Pradesh. In Assam, a jungle in 1840, by 1900 had 1.62m hectares under cultivation, especially in tea plantations. In all, the amount of irrigated land multiplied by a factor of eight. Historian David Gilmour states British colonial government had built irrigation network with Ganges canal and that, "by the end of the century the new network of canals in the Punjab" were in place.

Much of the increase in irrigation during British colonial era was targeted at dedicated poppy and opium farms in India, for exports to China. Poppy cultivation by the British Raj required reliable, dedicated irrigation system. Large portions of the eastern and northern regions of India, namely United Provinces, Northwestern Provinces, Oudh, Behar, Bengal and Rewa were irrigated to ensure reliable supply of poppy and opium for China. By 1850, the Asian opium trade created nearly 1,000 square kilometers of poppy farms in India in its fertile Ganges plains, which increased to over 200.000 hectares by 1900. This diversion of food crop land to cash crop use, state scholars, led to massive famines over the 1850 to 1905 period.

Major irrigation canals were built after millions of people died each in a series of major famines in the 19th century in British India. In 1900, British India (including Bangladesh and Pakistan) had about 13 million ha under irrigation. In 1901 the Viceroy, Lord Curzon, appointed a Commission chaired by Sir Colin Scott-Moncrieff to draw up a comprehensive irrigation plan for India. In 1903 the Commission's report recommended irrigation of an additional 2.6 million hectares. By 1947, the irrigated area had increased to about 22 million ha. In Northwestern British India region alone, with the colonial government's effort, 2.2 million hectares of previously barren land was irrigated by the 1940s, most of which is now part of Pakistan. Arthur Cotton led some irrigation canal projects in the Deccan peninsula, and landmarks are named after him in Andhra Pradesh and Tamil Nadu. However, much of the added irrigation capacity during the colonial era was provided by groundwater wells and tanks, operated manually.

IRRIGATION TRENDS SINCE 1947

India's irrigation covered crop area was about 22.6 million hectares in 1951, and it increased to a potential of 90 mha at the end of 1995, inclusive of canals and groundwater wells. However, the potential irrigation relies on reliable supply of electricity for water pumps and maintenance, and the net irrigated land has been considerably short. According to 2001/2002 Agriculture census, only 58.13 million hectares of land was actually irrigated in India. The total arable land in India is 160 million hectares (395 million acres). According to the World Bank, only about 35% of total agricultural land in India was reliably irrigated in 2010.

The ultimate sustainable irrigation potential of India has been estimated in 1991 United Nations' FAO report to be 139.5 million hectares, comprising 58.5 mha from major and medium river-fed irrigation canal schemes, 15 mha from minor irrigation canal schemes, and 66 mha from groundwater well fed irrigation.

India's irrigation is mostly groundwater well based. At 39 million hectares (67% of its total irrigation), India has the world's largest groundwater well equipped irrigation system (China with 19 mha is second, USA with 17 mha is third).

India has spent Rs. 16,590 crore on irrigation development between 1950 and 1985. Between 2000-2005 and 2005-2010, India proposed to invest a sum of Rs. 1,03,315 crore (INR) and Rs. 2,10,326 crore (INR) on irrigation and flood control in India.

Irrigation Potential and Sources of Irrigation

The principal sources of irrigation in India can be divided into the following:

(i) canals

(ii), wells

(iii) tanks and

(iv) others.

Approximately 31% of the irrigated areas in India is watered by canals. This includes large areas of land in Punjab, Haryana, Uttar Pradesh, Bihar and parts of southern states. Wells are now spread over large areas of Punjab, Uttar Pradesh, Bihar, Rajasthan and Tamil Nadu. Tanks are constructed for storing water in rainy season which is subsequently used for irrigation] purposes.

Development of irrigation and Government Policy during the Plan Period:

At the beginning of planning in India in 1950-51 irrigation schemes were divided into three categories – Major, costing more than Rs 5 crore each; medium costing individually between Rs. 10 lakh Rs. 5 crore each; and minor, costing less than Rs 10 lakh each. A new classification was adopted in April 1978.1 According to this classification, major schemes as those) having CCA (Culturable Command Area) between 2,000 hectares and 10,000 hectares, and minor schemes as those having CCA of less than 200 hectares.

Massive investment was made on irrigation during the planning period. From Rs 455 crore in the First] Plan, the expenditure on irrigation rose to Rs 36,649 crore in the Eighth Plan. The outlay for irrigation an (flood control in Nineth Plan was Rs 63,682 crore. Of this, Rs. 48,259 crore (76%) was for major and medium irrigation projects and the balance Rs. 15,423 crore (24%) for minor irrigation.

Common Area Development (CAD) programme and flood control. The outlay on irrigation and flood control in the Tenth Plan (2002- 07) had been kept at Rs. 1, 03,315 crore. Substantial expenditure has gone in for developing the major and medium irrigation potential specially the major river valley project like the Bhakra Nangal Project (Punjab), Beas Projects (Punjab and Haryana), Hirakund Dan Project (Orissa), Damodar Valley Corporation (Bihar and West Bengal), Nagarjuna Sagar Project (Andhra Pradesh and Karnataka), etc. However, minor irrigation continues to occupy an important place as its share in the total irrigation potential.

Introduction of the new agricultural strategy an (high yielding varieties programme required substantial irrigation facilities. Consequently in the Fourth Five Year Plan new schemes were introduced. Fifth Five Year Plan also introduced a comprehensive programme and 38 command area development authorities were set up covering 50 irrigation projects.

Rural Infrastructure Development Fund (RIDF) was launched to provide loans to the State Government for financing rural infrastructure projects including soil conservation and watershed management etc. In 1996-97, programme called "Accelerated Irrigation Benefit Programme" (AIBP) was launched by the Government of India.

Under this programme, The Centre provides additional assistance by way of loans to the states on matching basis for early completion of selected large irrigation and multi-purpose projects. The Ninth Plan noted that irrigated area, which constitutes only about 40 per cent of the total cultivated area, contributes almost 60 per cent of the food-grain production in the country. This shows the contribution of irrigation to agricultural output.

Limitations of Irrigation on Account of Certain Problems Related to Irrigation:

Despite large-scale investment and expansion of irrigation facilities, it is a matter of serious concern that about 60 per cent of the total cropped area is still dependent on rain. There are a number of problems related to irrigation and they have to be solved.

(1) Delays in completion of projects:

The biggest problem in our major and medium irrigation sectors right from the First Five Year Plan has been the tendency to start more and more new projects resulting in wanton proliferation of projects. There is also delay in utilisation of potentials already present. In most of the projects, there have been delays in construction of field channels and water courses, land leveling and land shaping.

(2) Inter-state water disputes:

Irrigation is a state subject in India. Development of water resource is, therefore, being planned by states individually taking into account their own needs and requirement. However, all major rivers are inter-state in character. As a result, differences with regard to storage, priorities and use of water arise between different states. Narrow regional outlook brings inter-state rivalries over distribution of water supply.

(3) Regional disparities in irrigation development:

The Ninth Five Year Plan Document estimated that the water resource development in North Eastern region through major, medium and minor schemes is only at the level of 28.6 per cent whereas in the Northern region it has reached about 95.3 per cent. This indicates a wide regional variation in the development of irrigation facilities.

(4) Waterlogging and salinity:

Introduction of irrigation has led to the problem of waterlogging and salinity in some of the states. The working group constituted by the Ministry of Water Resources in 1991 estimated that about 2.46 million hectares in irrigated commands suffered from water logging. The working group also estimated that 3.30 million hectares had been affected by salinity/alkalinity in the irrigated commands.

(5) Increasing cost of irrigation:

The cost of providing irrigation has been increasing over the years from the First Five Year Plan to Tenth First Five Year Plan.

(6) Losses in operating irrigation projects:

While just prior to Independence (1945-46) public irrigation schemes showed a surplus after meeting working expenses and other charges. The position deteriorated considerably in the post-Independence period.

(7) Decline in water table:

There has been a steady decline in water table in the recent period in several parts of the country, especially in the western dry region, on account over exploitation of ground water and insufficient recharge from rain water.

THE RATIONALE OF INTERLINKING RIVERS

The availability of water resources in various river basins of the country is highly uneven. While 32% of the total water resources are still available in the Brahmaputra basin, and 28% of the total water resources in the Ganga basin, this availability is merely 0.2% in the Sabarmati basin. The water scarcity in river basins is growing fast with increase in population. Based on this criteria and availability of water in different river basins, some basins have already have scarce water resources and many more basins are likely to have water scarcity with the growing population by year 2025.

Out of 12 major and 48 medium river basins in India, the government predicts that by 2025 the deficit river basins will be Ganga, Subernarekha, Krishna, Mahi, Tapi, Cauvery, Pennar and Sabarmati. The surplus basins would be Brhamaputra, Barak, Narmada, Brahmani-Baitarani, Mahanadi, Godavari and Indus. Considering the precautionary approach to face the challenge of water scarcity in several river basins in 2025, the government has taken the interlinking of rivers as its definitive answer.

India receives an annual precipitation of 4,000 billion cubic meter (BCM, equivalent to 4,000 cubic kilometers) of which 75% occurs just in the four months of the monsoon period. From the annual precipitation, 1,869 BCM of water appears as runoff in various river basins. The utilizable water resource has been assessed as 1,132 BCM. Rainfall in India is erratic and uneven that ranges from 11,000 millimeter annually in some parts of North Eastern India to 100 millimeter in Western India. To address this climatic disparity and find ways for augmentation of utilizable water, interlinking of rivers has been put forward by Indian government. The ILR proposal also promises to enhance the production of food grains up to 380 million tonnes to help meet ever increasing population demand by 2025. The ILR is projected to provide 35 million hectares of additional potential arable land and 34,000 megawatt of electricity.

The idea of linking rivers for various purposes in the sub-continent is not new. Sir Arthur Cotton conceived a plan to link rivers in Southern India for inland navigation in the nineteenth century. While the project was partially implemented, the river-linking canals could not survive in the face of rapid development of railways. The idea of a Ganga-Cauvery Link was proposed by Dr. K.L. Rao, former Union Minister for Irrigation, in 1972. The link involved a lift of water 450 meters from the flood flows of the Ganga, withdrawing 60,000 cusecs (60,000 cubic feet per second) of water for 150 days in a year including a 2,640 kilometer long link canal. The plan was discarded as it involved an high cost ($2.7 billion) and required a large energy consumption to operate its pumps. Subsequently, Capt. Dinshaw J. Dastur, an aviator, advanced a proposal for the "Garland Canal" system that consisted of two canals: (1) the Himalayan Canal and (2) the Central and Southern Garland Canal. The government agency indicated that the project was impracticable, technically unsound, and economically prohibitive.

Development of Irrigation in India

Water is to land what food is to human body. The part played by irrigation in Indian agriculture is all the greater due to the uncertainty, unequal distribution, as well as the insufficiency of rainfall. In large parts of the country, rainfall is the only source of water-supply and its failure causes almost famine conditions.

That is why "Irrigation works... although almost un-known in Northern Europe, have existed in India...from time immemorial."

The oldest and the most famous was the Cauvery delta system of canals in Madras which Sir Arthur Cotton strengthened and improved in 1835-1836, and the success of which encouraged him to propose the great Godavari works which he afterwards constructed. Most of the weirs on the Tungbhadra were constructed by the Hindu King, Krishna Raya, at the beginning of the 16th century.

In Northern India, the Mahommadans made frequent attempts to utilise the water of the river Jamuna. The western Jamuna Canal was started by Feroz Shah Tuglak in the middle of the 14th Century while the Eastern Jamuna Canal was made at the beginning of the 18th Century.

In the Punjab, the small Hash canal was constructed by early rulers for carrying water from the river Ravi to Lahore and Amritsar on very much the same alignment as the present Upper Bari Doab. Some of these early works were destroyed by internal wars before the establishment of the British rule, and the number and utility of those which were left was much diminished by want of repairs.

Although, the earliest results attained by direct govt. agency on the Cauvery, Godavari and Jamuna works were sufficiently encouraging, the East India Company was reluctant to commit itself to any extensive scheme of state irrigation works.

In the words of Sir Arthur Cotton, the policy of the company was to "do nothing, have nothing done, let nobody do anything. Bear any loss. Let the people die of famine, let hundred of lakhs be lost in revenue for want of water… rather than do anything."

However, the series of devastating famines which visited the country at regular intervals soon forced the company to give attention to the question of providing irrigation facilities. A General Superintendent of Irrigation was appointed in 1823 and the repair and improvement of the old works was taken in hand. The Western Jamuna Canal was re-opened in 1821 and the Eastern Jamuna in 1830.

In 1842, the construction of Ganges Canal was begun but, owing to delays caused by the Afghan and Sikh wars, it was not opened till 1854. Another of the great works of Northern India undertaken by the East India Company was the Upper Bari Doab. The construction of this canal was undertaken soon after the annexation of the Punjab but was not opened until 1859.

While these works were being constructed in Northern India, the Madras Engineers were engaged in the development and re-modelling of the old irrigation works in the delta of the Cauvery river. The success of this scheme led Sir Arthur Cotton to propose a similar scheme on the Godavari river which was sanctioned in 1846 while the work on the river Krishna was Completed in 1855.

Apart from these larger works, various smaller schemes were also undertaken under orders of the East India Company and these included the construction of small canals in the Dehradun and Rohilkhand area of U.P. and the repair of various tanks in South India.

All these irrigation schemes were carried out of revenue surpluses and, though profitable, their construction imposed a strain on the resources of the company. It was, therefore, suggested that the construction of large irrigation works, like that of railways, could be most advantageously entrusted to private enterprise.

The real reason, however, lay in the fact that, after the heyday of the Industrial Revolution, safe and profitable investment outlets had narrowed in England while railway investments in territories outside Britain's political control had become risky. The Government, therefore, obliged by providing a 5% guarantee on investment in irrigation undertaken by private British companies.

Two companies were floated in England to take up these ambitious schemes, section by section. The first company, known as the East India Irrigation and Canal Company, was formed in 1858 with the intention of constructing irrigation and navigation canals in Orissa. The company soon got into difficulties; the govt. began by giving a subsidy and ended by taking over the concern in 1869 at an exorbitant rate.

The second company, known as the Madras Irrigation Company, was formed in 1863. It rarely succeeded in meeting its working expenses. In addition to the guaranteed interest of 5%, a large sum was advanced to the company free of interest and, ultimately, it was taken over by the govt. in 1882.

In the light of this experience, it was realised that the construction and control of such irrigation works could not be conveniently entrusted to private enterprise. Thus was initiated, under the viceroyalty of Lord Lawrence, a new policy under which works that gave a fair promise of proving directly remunerative, were to be constructed by the state from loan funds.

State initiative and control gave irrigation projects the benefits of proper financial aid and well-coordinated technical knowledge and experience the lack of which had led to the failure of state-aided private companies.

Between 1862—1882, five big 'Loans works', namely, the Sir hind Canal (Punjab), the Lower Ganges and the Agra Canal (U.P.), the Lower Swat Project (N.W.F.P, now in Pakistan) and the Dam at Kharagvasla near Poona were constructed by the govt.

In all, a total of £ 12 million was spent on irrigation works up to 1880, while the total irrigated area amounted to 29 million acres. It may be noted that in the case of Bombay Deccan, Berar and Central provinces ... areas prone to frequent famine, the irrigated area formed an insignificant proportion of the total cultivated area.

The provinces which were better served by irrigation owed the most important part of their large irrigation works to the period before 1860. Thus, the four important works of North India, the Upper Bari Doab in the Punjab, the two Jamuna Canals and the magnificent Ganges Canal, were all constructed before the Crown took over the administration of the country.

Similarly, the irrigation works in the Deltas of the Krishna, Cauvery and Godavari were constructed before 1858. The progress made in the construction of irrigation works between 1859—80 was thus extremely inadequate and unsatisfactory. Two factors were responsible for this.

These were, firstly, the unfortunate belief that the key to the solution of the famine problem lay more in the extension of railways than in the development of irrigation and, secondly, the financial difficulties of the govt. which stood in the way of its raising sufficient funds for investment in irrigation works.

The terrible Madras Famine of 1877, however, gave Sir Arthur Cotton an opportunity to rouse the attention of the British public to the un-wisdom of neglecting irrigation in India.

The famine commission of 1880, drawing pointed attention to the importance of irrigation, remarked that "among the means that may be adopted for giving India direct protection from famines arising from drought, the first place must unquestionably be assigned to works of irrigation."

The commission further laid down that the value of irrigation works was to be judged first by "the direct protection afforded by them in years of drought" and, secondly, by the great services they rendered in imparting "certainty to all agricultural operations, increasing the out-turn per acre of the crops, and enabling more valuable crops to be grown."

Following these recommendations, a number of canal irrigation schemes were pushed forward in the Punjab, Bengal, Bombay and Sind. In the Punjab, the extension of the western Jamuna Canal and the construction of the Chenab, Lower Sohag, Para Inundation Canal and the Sidnai Canal was sanctioned. In Bengal, the completion of the Orissa Canal system was undertaken. The Betwa Canal in U.P. was the first protective work to be completed out of the newly created Famine Relief and Insurance Fund. Other protective works completed by 1900 included the famous Nira and Periyar Gokak, and Mhawad canal system in the Deccan and the Jamrao and Western Nara canals in Sind.

The Sarda Canal in Oudh could not be sanctioned as "more urgent irrigation works were needed elsewhere." The overall result of these irrigation works, taken together with protective works, was that the canal-irrigated area in India increased from 10.6 million acres in 1878—79 to 15.3 million acres in 1896—97-an increase of 50%. This increase, though satisfactory in itself, fell far short of the needs of the country or even of what could have been achieved had the govt. given the same attention to irrigation as it bestowed upon the railways. Even after the emphatic declaration of the Famine Commission in favour of irrigation, the first place continued to be given to the railways.

For example, up to the end of 1902—3, the total expenditure by the govt. on major and minor works of irrigation amounted to nearly Rs. 43 crores. In contrast, the total outlay on the state and guaranteed railways had amounted, by 30 June 1905, to Rs, 359 crores.

This 'unwise extravagance' in regard to railways and 'unwise niggardliness' in relation to irrigation cannot be explained by the financial considerations of net return on investment or the paucity of funds.

As the Famine Commission of 1898 concluded, "Even apart from their value in increasing the wealth of the country in ordinary years and in preventing or mitigating famines in years of drought" the results of irrigation works "regarded merely from the direct financial return on the money invested was a great advantage, to the State."

The major irrigation works, on an average, yielded a net revenue of 7% and loans were available in London at 3.4%. On the other hand, the large majority of railway lines, with the exception of a few, were a drain on the exchequer, the average loss sustained by the govt. being an estimated Rs. 1 crore per year. Obviously, financial considerations could not have been the basis of the govt's preference for railways.

Similarly, Dr. Anstey's argument that the limitations of knowledge and skill and the fears of a possible exhaustion of river water were among the factors responsible for restricting in investment in irrigation cannot be accepted because by 1900, India had utilised only a fraction of her river-water potential while engineers like Arthur Cotton were available who were willing and able to face the technical challenge involved in the construction of irrigation works.

It may also be argued that further investment in irrigation works before the First World War would have swollen total expenditure and, therefore, aggravated the problem of rise in prices. But even this was not a compelling argument for two reasons.

In the first place, the stimulus for the rise in the prices came, to a considerable extent, through rise in the prices of Indian exports and an increase in expenditure could not have very much affected export prices.

In the second place, a large part of the increased expenditure would have leaked abroad in the form of increased imports of capital goods and in the form of salaries of technical personnel. R.C. Dutt thinks that English men made a 'geographical mistake' in that "the British nation, more familiar with railways than canals in their country, did not adequately realise the great importance of irrigation in India," and, therefore, did not expand irrigation with the same keenness with which railways were multiplied.

The common man in Britain might have been ignorant about Indian conditions, but the British administrators were certainly not. The truth is that the public works policy of the govt. was influenced by 'Imperial and commercial' considerations. Railways benefitted the trade—import of manufactured goods and export of raw-materials....while canals helped agriculture in which the English trader had little interest.

That is why the govt. gave priority to railways over canals and men like Arthur Cotton and Lawrence could not make any headway against the general and 'deadening indifference of the govt.'

Besides, being based on commercial considerations, the policy of the govt. was defective in that the choice of irrigation schemes was made on the financial consideration of whether they would pay. Protective works could not be financed out of current revenues or borrowed funds.

The Famine Insurance Fund, which was the only source of finance for such works, was too meagre to 'support any large programme of protective works.' The irrigation policy of the govt., therefore, bore no relation, till the end of the 19th century, to the needs of the situation.

It was only after 1900, when an enormous network of railways had been laid for meeting Britain's trade requirements, that funds as well as attention began to be spared for irrigation.

The appointment of the Irrigation Commission in 1901 was, therefore, not merely an indication of govt's, readiness to devote greater attention to irrigation but it also marked a fundamental change in the thinking of the govt. on the purpose of irrigation works in India. It did not mean stopping further extension of productive works which the govt. constructed 'as a matter of business.'

It only meant the beginning of a move in the direction of providing irrigation facilities to those areas "where no works of productive class were possible and which must be protected if at all, at a certain cost to the revenues."

The commission pointed out that irrigation works has not, in the past, received the attention that they deserved, that there was still great scope for the extension of remunerative irrigation works in some parts of the country, that there had been over the past 25 years very little increase in the area covered by minor irrigation and that besides extension of state protective works, much could be achieved by construction of private works such as wells through state advances and revenue concessions.

A 20 years plan, involving a total expenditure of Rs. 44 crores, was drawn up by the commission. Of this sum, Rs. 15 crores, were proposed for investment on productive works, Rs. 20 crores on protective works and rupees 9'crores on intermediate works.

Following the recommendations of the commission, the Govt. began to pay greater attention to irrigation. Formerly, greater portion of famine grant was set apart for railways, but now it was decided to allocate the whole of this sum to protective works. In 1910, this sum was found to be insufficient and a further annual contribution, not exceeding Rs. 25 lakhs, from the revenues was sanctioned for this purpose.

The chief work completed before the 1st World War was the Triple Canal Project in the Punjab. (Upper Jhelum, Upper Chenab and Lower Bari Doab) Work on the Lower Jhelum Canal was also completed although the scheme had been sanctioned in 1888. Most of the other work carried out were famine protective storage works.

These included the Mahanadi, Wainganga, Tendula and Ram Tek Canal projects in the central provinces and the Ken, Dhosan, Ghaghar, and the Gorai Canals in U.P., and the Bhandaradara Dam and the Bhatgarh Dams in the Bombay Deccan. The Total capital invested in irrigation works rose from Rs. 42 crores in 1900 to Rs. 78 crores in 1920.

The end of the war saw the world entering upon a period of unprecedented prosperity. Wages rose and so did prices and, with these, the value of land and food grains also went very high.

Fortunes were made out of land, more particularly where irrigation made it possible to grow superior crops. The result was that projects which had previously appeared doubtful from the financial point of view now offered prospects of high returns. Besides, under the reforms of 1919, irrigation became a reserved provincial subject and the new provincial govts. were keen to push forward irrigation schemes. What is more, accepting the recommendations of the Irrigation Commission, the distinction between 'productive' and 'protective' works was also abolished in 1922. Now all works could be financed out of loans.

The stage was, therefore, set for a more rapid expansion of irrigation facilities in the country. Certain large projects were hurriedly prepared and sanctioned in quick succession. These were the Sukkur Barrage scheme in Sind, the Sutlej Valley scheme to irrigate part of the Punjab, Bikaner and Bahawalpur, and the Sarda Canal Project in U.P.

These were all first-rate though controversial schemes and their total cost was almost equal to the total expenditure incurred till then on both productive and unproductive works. Other works completed were the Cauvery Mettur Project, The Nira Right Bank Canal and the Damodar Canal.

Another notable development of the period was the establishment, in 1931, of the Central Board of Irrigation for the purpose of coordinating research in irrigation.

Although the II World War left very little time and resources for the construction of any large works, still certain new projects, especially the Tungbhadra project, were taken in hand. During the period of the war were also established the Central Water-power Irrigation and Navigation Commission and the Central Ground Water Organisation.

The first was required, among other things, to coordinate schemes of irrigation, water-power generation, navigation, and, if so required for the construction of new schemes. The second was meant to conduct research in the utilisation of underground water.

The partition of the country was a severe blow to India in so far as irrigation facilities were concerned. It saddled India with 82% of the population but only 69% of the total irrigated area of the undivided country, many major irrigation works having been left in west Punjab and Sind.

The gravity of the situation can be realised from the fact that in Pakistan, 45.2% of the net sown area was irrigated while the corresponding percentage in Indian Union was only 18.9%.

It was only after Independence that irrigation began to get the attention it really deserves in an agricultural country. It was only now fully understood that "large-scale development of irrigation and power helps to re-build the agricultural economy and pave the way for the rapid industrialisation of the country."

An ambitious programme of multipurpose projects was undertaken for a planned extension of irrigation facilities in the country. The First Plan mainly included schemes which had already been taken up in the period prior to the commencement of planning. In the Second Plan, there was distinct shift towards medium schemes.

However, since a large number of schemes continued to be under construction even at the end of the second plan, the emphasis during the Third Plan was on completion of continuing schemes. As such, very few new schemes were taken up in the Third Plan. All told, during these three plans, a total investment of Rs. 1300 crores was made on the construction of 295 major and minor irrigation works.

As a result, total irrigated area increased from 56 million acres before 1950—51 to 87 million acres in 1965-66. Major schemes completed included the Bhakra Nagal, Damodar Valley, Hirakud, Kosi, Koyna, Krishna, Chambal, Rihand, Tungbhadra, Mayurakshi and Lower Bhowani.

Though the efforts of the govt. were concentrated on the construction of canals, yet other means of irrigation were not altogether ignored. From times immemorial, wells had been tried as a method of irrigation in the Punjab, U.P., Madras, and parts of Bombay.

These were privately owned and the views of the govt. were well summed up by the Agricultural Commission when it said that "the construction of wells is essentially a matter of private enterprise."

Amazingly, the area under well irrigation remained stationary for a very long period despite the encouragement given by the govt. through loans under the Land Improvement Act. In 1880 the Famine Commission estimated the total well-irrigated area at 10.8 million acres; the Irrigation Commission (1901) put the figure at not less than 12.8 million acres. It was 11.7 million acres in 1925—26 and 11.37 million acres in 1933.

The reason probably is that with the extension of canal irrigation in several parts of the country, many old wells became useless and were no more used. However, after the Bengal famine, greater official encouragement began to be given to the construction and repair of wells.

Under the Grow More Food campaign and subsequently the plans, liberal help was given in the shape of loans to encourage cultivators to construct or repair their own wells. As a result, the area irrigated by wells rose to 18.2 million acres in 1960—61.

The Ganges valley Tube well Irrigation Scheme marked a new development in the practice of well irrigation. Under that scheme, 1300 tube-wells were constructed during the three years 1935—37. Since then, the number of tube wells went on rising. At the end of the Third Plan, the number of tube-wells in the country had risen to 80,000.

Tank irrigation is conspicuous by its absence in the Punjab but was of considerable importance in Madras, Deccan, Mysore, and Bihar. The reasons for the concentration of tanks in the South are obvious. Firstly, the rivers in the South are not snow-fed and depend solely on rainfall.

Secondly, there are many streams which cause floods during the monsoons but later dry up. Their water can be stored and floods prevented through construction of tanks. The land is uneven and its bed rocky, making it very difficult to construct canals or dig wells.

While the smaller tanks were constructed and maintained by private enterprise, the larger ones were always maintained as state irrigation works. These irrigated over 8 million acres in 1900. After the report of the Irrigation Commission, the Govt. began to make greater effort to renovate and repair existing tanks so as to make them serviceable.

The Bombay Govt. appointed a special officer to look after the 'Bhandaras' in Nasik and Khandesh districts. The area irrigated by tanks, therefore, rose to about 11.5 million acres in 1960—61.

"These Irrigation works made for security of life, increased the yields and the value of lands and the revenue derived from it. They lessened the cost of famine relief and helped to civilize whole regions. In addition, they yielded a handsome profit to the govt." But irrigation was not always an un-mixed blessing.

In the absence of a very good drainage system, water-logging and Malaria often followed. In U.P. and the Punjab, water logging often led to the rising to the surface of a saline substance called 'Kalar' which made the land almost un- cultivable.

TWO COMPONENTS: HIMALAYAN RIVERS AND PENNINSULAR RIVERS

In view of the K L Rao's proposal, the Indian Ministry of Water Resources in 1980 framed a National Perspective Plan (NPP) for Water Development and the National Water Development Agency (NWDA) was established in 1982, to carry out studies in the context of the National Perspective. The NPP has two components: a) Himalayan Rivers and b) Peninsular Rivers. NPP has proposed 30 river links involving 37 rivers. As the proposal includes the Himalayan Rivers, which are transboundary in nature, the other South Asian riparian countries have conveyed apprehensions to India.

The Himalayan Rivers development component ensures construction of storage reservoirs on the principal tributaries of Ganga and Brahmaputra rivers in India, Nepal and Bhutan along with interlinking of river systems to transfer surplus flows of the eastern tributaries of the river Ganga to the west, apart from linking of the main Brahmaputra and its tributaries with Ganga and Ganga with the river Mahanadi.

The Himalayan Rivers component carries 14 links, of which 12 interdependent links and 2 independent links. The links in Himalayan segment consists of some within the Ganga system (Kosi-Ghagra (Karnali), Gandak-Ganga, Ghagra-Yamuna, Sarda-Yamuna, some links between neighbouring rivers in the Brahmaputra system (Manas-Sankosh-Teesta); a couple of links between those two systems (Teesta-Ganga or an alternative Brahmaputra-Ganga link); one long link from Sarda to Sabarmati through the Yamuna and Rajasthan; one from the Ganga to Subernarekha via Damodar and then on to the Mahanadi.

The average flood discharge of Ganga is 50,000 cubic meter per second and for the Brahmaputra the average flood discharge is 60,000 cubic meter per second. Through these links the amount of water to be diverted are 1,500 cusec in Brahamaputra and 1,000 cusecs in Ganga. However, the National Commission for Integrated Water Resource Development Plan in its September 1999 report did not discuss the proposed Himalayan links in detail because the data are classified as confidential, but did observe that the costs involved and the environmental problems would be enormous. The Himalayan links are not to be touched now as there are international dimensions including Nepal and Bangladesh concerns.

The Peninsular rivers development component is divided into four major parts: a) interlinking of Mahanadi-Godavari-Krishna-Cauvery Rivers basins; b) interlinking of west flowing rivers, north of Bombay and south of Tapi; c) interlinking of Ken-Chambal and d) diversion of other west flowing rivers. There are 16 links are proposed by the National Water Development Agency (NWDA), out of which 9 interdependent and 7 independent links are being proposed. Among the 16 links Ken-Betwa links has moved a step ahead.

In the proposal of NPP, the transfer of water has been proposed mostly by gravity; building of dams and storage, construction of canals and pumping of water where necessary (confined to around 120 meter). Pumping water over the Vindhya Mountains can transfer the Ganga-Brahmaputra water and its tributaries to regions in the south. The Ganga-Brahmaputra floodplains are about ten meters above mean sea level (MSL). The Vindhya Mountains are about 300 meters above MSL, separating the floodplains of the north from the Deccan Plateau, which is 250 meters above MSL.

The expected links where pumping would be necessary is Ganga-Subarnrekha (60 meters), Subarnarekha-Mahanadi (48 meters) and Godavari-Krishna (112 meters). The expected lift would be 1,200 cusecs water over 116 meter from Mahanadi to the Krishna basin. Some believe that the electric power required to pump water to such heights will be close to the current power generation of the entire nation, but as the table below indicates, this may not be at all the case. The Interlinking Rivers proposal carries the structure and design of the canals with 1:3,000 to 1:5,000 slope or 0.33 to 0.20 meter per kilometer. Expected maximum flow velocity of water transfer is 2 meters per second.

- Project classification
- Irrigation Projects in India are classified as follows:
- Major Irrigation Projects
- Medium Irrigation Projects
- Minor Irrigation Projects
- Micro-irrigation Projects

Since 1950, irrigation works were classified on the basis of cost incurred for the projects' implementation, governing and dissemination. However, the Planning Commission of India adopted the classification of projects on the basis of culturable command area (CCA).

MINOR IRRIGATION PROJECTS

Minor irrigation project is a classification of irrigation projects used in India. A project with a designed to irrigate an area of 2000 hectares or less is classified as a minor irrigation Before the Fifth Five-Year Plan, irrigation schemes were classified on the basis of investments needed to implement the scheme. Since the Fifth Five-Year Plan, India has adopted the command area-based system of classification.

Micro Irrigation Projects

Micro Irrigation Fund (MIF) of INR 5,000 crore was established "to bring more land area under micro-irrigation as part of its objective to boost agriculture production and farmers income", NABARD offers low interest rate to state govts "to promote micro-irrigation, which currently has a coverage of only 10 million hectares as against the potential of 70 million hectares." Drip irrigationis promoted.

The Initiative's Recent Momentum

The ILR proposal has backed by the three organs of the Indian political system - executive, legislature and judiciary in a strange manner. For the last two decades, the National Water Development Agency (NWDA)" had been working on the subject but its proposals were non-starters for various reasons. Some believe that the "non-project" has suddenly become the most important undertaking of the Indian government.

However, can argue that during 2001 and early 2002, the ground situation of severe drought and floods in several states had compelled the government to re-open the ILR project proposal. occurred in a cascading manner to put forward the ambitious project, hitherto a dormant idea. The President's national address on August 14, 2002, the file of Public Interest Litigation (PIL) to clean up Yamuna River in Delhi in late August, and a hearing on the Cauvery dispute by India's Supreme Court on October 31, 2002 led the union government to issue a statement in Parliament in December for interlinking of rivers. The Prime Minister subsequently announced the setting up of a Task Force on ILR to consider the modalities of implementing the project.

Constraints and Concerns

An opinion poll conducted by the news weekly India Today prior to the 14th General Elections 2004, showed that 78 per cent of voters spread over 185 Parliamentary constituencies supported the project. Does that mean there is a consensus on ILR? The very idea of inter-linking of rivers as a flood control measure is being discarded by hydrologists. Moreover, The ILR is no answer at all to the needs of areas unserved by rivers. For drought mitigation there are enough alternatives available such as traditional rainwater harvesting, moisture conservation, tank-rehabilitation etc. The water experts believe that the primary answer to drought has to be local; it is only thereafter, and in some very unpromising places where no other options are feasible, that the bringing in of some external water may need to be considered.

Some argued that the interlinking project has the potential for generating conflicts within the state, among the state and among countries in the cooperative management of the project in international river basins. Major river basins in India are ridden with conflicts over sharing of water resources. Provinces are hesitant to divert surplus water even they show it as deficit. In NWDA's own admission that the major constraints for implementation of the ILR will be international dimension in several links including construction of dams in upstream Nepal and Bhutan. At the other end, Bangladesh, the lower riparian country has officially conveyed its apprehension about the Indian ILR project.

The unilateral nature of the Himalayan component of the proposed ILR project is the most worrisome aspect for the downstream riparian Bangladesh and upper-stream riparian Nepal, who are afraid of the environmental damages resulting from large-scale inter-basin transfer of water in the Ganga-Brahmaputra-Meghna region. Other problems are submergence of large tracts of forests and large numbers of displacements. One conservative estimate says that the ILR will submerge 1,675,000 hectares of forests and agriculture land including 1,050,000 ha for reservoirs. The ILR includes 60 new reservoirs. While one government calculation predicts of 0.45 million people would be displaced, others estimate the figure would be 3.47 million.

India's most ambitious project is being carried out with utmost caution. The public mood is not evenly divided as the information on ILR is very limited and is not under the layman's purview. The polarization between the pro-ILR and anti-ILR has been widening to further make it difficult for the government to complete it within a stipulated time frame given by the apex court of India. Given the peculiar and intricate federal structure of India with water coming under provincial government's power, it is difficult to garner consensus on the viability of the project. Similarly, to get Union environmental clearance may pose a hurdle to the ILR. To redesign a natural river course is undoubtedly within the control of a scientific human. But its cascading impact is far reaching, in some aspects possibly uncontrollable.

Irrigation in India

Irrigation in India refers to the supply of water from Indian rivers, tanks, wells, canals and other artificial projects for the purpose of cultivation and agricultural activities. In country such as India, 64% of cultivated land is dependent on monsoons. The economic significance of irrigation in India is namely, to reduce over dependence on monsoons, advanced agricultural productivity, bringing more land under cultivation, reducing instability in output levels, creation of job opportunities, electricity and transport facilities, control of floods and prevention of droughts.

Potential capacity

India has an irrigation potential of 139.89 million hectares, out of which a minimal 108.2 million hectares(77.35%) of the total land that can be irrigated has been utilized. Currently, about 30% of the net cultivated area has benefited from the irrigation projects that have been implemented.

A sum of 16,590 crore has been spent of irrigation development up to the 7thFive-Year plans of India. The 10th and 11th Five Year Plan have proposed to invest a sum of 1,03,315 crore and 2,10,326 crore on irrigation and flood control in India.

Project classification

Irrigation Projects in India are classified on three major aspects into:

1. Minor Irrigation Projects
2. Medium Irrigation Projects
3. Major Irrigation Projects Since 1950, irrigation works were classified on the basis of cost incurred for the projects' implementation, governing and dissemination., However, the Planning Commission of India adopted the classification of projects on the basis of culturable command area(CCA).

Water is very important for survival of all forms of life- plant as well as animal. India, by virtue of its peculiar placement in the foothills of the Himalayas and the Deccan Plateau running through it, has vast water resources which have been very meagrely tapped. Conventional and recognised means of irrigation are tanks, wells and canals.

Wells: Well irrigation is an important type of irrigation in India. Wells are particularly suitable for small farms. The important well-irrigated States are Uttar Pradesh, Punjab, Tamil Nadu and Maharashtra. In these States water-table is high, soil is soft and, therefore, wells are easily sunk.

Tubewells are an important development in India. They are worked by electricity or diesel oil and thus, they relieve our cattle of much of the strain. They are being quickly developed in Uttar Pradesh, Bihar, Haryana and Punjab. This is because these have ample sub-soil water.

Wells and tubewells account for about 48 percent of the total irrigation in India.

Tanks: Tanks are also an important and ancient source of irrigation. They are of considerable importance in central and southern India, specially in Andhra Pradesh and Tamil Nadu. About 8 percent of the total irrigatedarea is irrigated by tanks.

Canals: Canals are the most important means of irrigation in the country. Some canals were constructed by the early Hindu and Mohammedan kings. Most of the canals, however, are the product of the British rule. At present, canals irrigate about 39 percent of total irrigated area of India. Most of the canals of the country are found in Uttar Pradesh and Punjab. Storage canals have been constructed in Deccan and Madhya Pradesh.

Major , Medium and Minor Irrigation Projects: The methods of irrigation used in India can be broadly classified into major, medium and minor irrigation schemes. Irrigation projects having Culturable Command Area (CCA) of more than 10,000 hectares each are classified as major projects. Those having a CCA between 2,000 hectares and 10,000 hectares fall under the category of medium irrigation projects. And theprojects which have a CCA of less than 2,000 hectares are classified as minor irrigation schemes. For the purpose of analysis the major and the medium irrigation projects are generally grouped together. Theseprojects comprise a network of dams, bunds, canals and other such schemes. Such projects require substantial financial outlay and are, therefore, constructed by the government or any other agency which may draw financial assistance form the government and financial institutions.

The minor irrigation projects, on the other hand, comprise all ground water development schemes such as dug wells, private shallow tubewells, deep public tubewells, and boring and deepening of dugewells, and small surface water development works such as storage tanks, lift irrigation projects, etc. Minor irrigation projectsor the groundwater development schemes are essentially people's programmes implemented primarily through individual and cooperative efforts with finances obtained mainly through institutional sources.

Irrigation Development

Creation for irrigation potential of 10 million hectares was targeted under Bharat Nirman during 2005-06 to 2008-09. The target was proposed to be met through completion of on-going major and medium irrigationprojects, and extension, renovation and modernization of existing projects. As per information provided by State Governments, the total irrigation potential created during the period is 7.31 million hectares against the target of 10 million hectares.

Some Irrigation and Multipurpose Projects

Bargi Project (Madhya Pradesh): It is a multipurpose project consisting of a masonry dam across Bargi river in the Jabalpur district and a left bank canal.

Beas Project (Joint venture of Haryana, Punjab and Rajasthan): It consists of Beas-Sutlej Link and Beas Dam at Pong.

Bhadra Project (Karnataka): A multipurpose project across the river Bhadra.

Bhakra Nangal Project (Joint project of Haryana, Punjab, and Rajasthan): India's biggest, multipurpose river valley project comprises a straight gravity dam across the Sutlej river at Bhakra, the Nangal dam, the Nangal hydel channel, two power houses at Bhakra dam and two power stations at Ganguwal and Kotla.

Bhima Project (Maharashtra): Comprises two dams, one on the Pawana river near Phagne in Pune districtand the other across the Krishna river near Ujjaini in Sholapur district.

Chambal Project (Joint project of Madhya Pradesh and Rajasthan): The project comprises Gandhi Sagar dam, Rana Pratap Sagar dam and jawahar Sagar dam.

Damodar Valley Project (West Bengal and Bihar): A multipurpose project for the unified development of irrigation, flood control and power generation in West Bengal and Bihar. It comprises multipurpose dams at Konar, Tilaiya, Maithon and Pancher; hydel power stations at Tilaiya, Konar, Maithon and Panchet; barrage at Durgapur; and thermal power houses at Bokaro, Chandrapura and Durgapur. The project is administrated by the Damodar Valley Corporation.

Dulhasti Power Project (Jammu & Kashmir): It is a 390 MW power project in Kishtwar region of Jammu & Kashmir on Chenab river. Work for this project started in 1981. The foundation stone was laid on April 15, 1983 by the then Prime Minister, Indira Gandhi. Work on this project was suspended due to threats of kidnapping and killings by Kashmiri militants resulting in long delay in completion of project.

Farakka Project (West Bengal): The project was taken up for the preservation and maintenance of Calcutta port and for improving the navigability of the Hoogly. It comprises a barrage at Jangipur across the Bhagirathi and a feeder channel taking off from the Ganga at Farakka and tailing into the Bhagirathi below the Jangipur barrage.

Gandak Project (Joint project of Bihar and Uttar Pradesh): Nepal also derives irrigation and power benefits form this project.

Ghataprabha Project (Karnataka): A project across Ghataprabha in Belgaum and Bijapur districts.

Hirakund (Odisha): World's longest dam, is located on the Mahanadi river.

Jayakwadi Project (Maharashtra): A masonry spillway across the river Godavari.

Kahalgaon Project (Bihar): The 840-MW Kahalgaon Super Thermal Power Project, a joint venture between National Thermal Power Corporation and the Russian State Enterprise Foreign Economic Association, was on August 12,1996 commissioned and put into commercial operation.

Kakrapara Project (Gujarat): On the Tapti river near Kakrapara, in Surat district.

Kangsabati Project (West Bengal): The project, put in operation in 1965, is located on the Kangsabati and Kumari rivers.

Karjan Project (Gujarat): A masonry dam across Karjan river near Jitgarh village in Nandoo Taluka of Bharuch district.

Kosi Project (Bihar): A multipurpose project, which serves Bihar and Nepal.

Koyna Project (Maharashtra): It is built on a tributary of river Krishna with a capacity of 880 MW. It feeds power to Mumbai-Pune industrial belt.

Krishna Project (Maharashtra): Dhom dam near Dhom village on Krishna and Kanhar dam near Kanhar village on Varna river in Satna district.

Kukadi Project (Maharashtra): Five independent storage dams, i.e. Yodgaon, Manikdohi, Dimbha, Wadaj and Pimpalgaon Jog. The canal system comprises (i) Kukadi left bank Canal, (ii) Dimbha left bank canal, (iii) Dimbha right bank canal, (iv) Meena feeder and (v) Meena branch.

Kundoh Project (Tamil Nadu): It is in Tamil Nadu whose initial capacity of 425 MW has since been expanded to 535 MW.

Let Bank Ghaghra Canal (Uttar Pradesh): A link channel taking off from the left bank of Ghaghra river of Girja barrage across Sarju.

Madhya Ganaga Canal (Uttar Pradesh): A barrage across Ganga in Bijnore district.

Mahanadi Delta Scheme (Odisha): The irrigation scheme will utilize releases from the Hirakud reservoir.

Mahanadi Reservoir Project (Madhya Pradesh): It has three phases: (1) Ravishankar Sagar Project and feeder canal system for supply of water of Bhilai Steel Plant and Sandur dam across Sandur village. (2) Extension of Mahanadi feeder canal. (3) Pairi dam.

Mahi Project (Gujarat): A two -phase project, one across the Mahi river near Wanakbori village and the other across Mahi river near Kadana.

Malaprabha Project (Karnataka): A dam across the Malaprabha in Belgaum district.

Mayurakshi Project (West Bengal): An irrigation and hydro-electric project comprise the Canada dam.

Minimato Bango Hasdeo Project (Madhya Pradesh): This project is locted at Hasdeo Bango river in Korba district and envisages construction of a masonry dam. A hydel power plant of 120 MW capacity has been commissioned on the Bango dam.

Nagarjunasagar (Andhra Pradesh): On the Krishna river near Nandikona village (about 44 km from Hyderabad).

Panam Project (Gujarat): A gravity masonry dam across Panam river near Keldezar village in Panchmahal district.

Parambikulam Aliyar (Joint venture of Tamil Nadu and Kerala): The integrated harnessing of eight rivers, six in the Annamalai Hills and two in the plains.

Pochampad (Andhra Pradesh): Across Godavari river.

Pong Dam (Punjab): It is an important hydro-electric project located on Beas river.

Rajasthan Canal (Indira Gandhi Canal- Rajasthan): The Project uses water released from Pong dam and provides irrigation facilities to the north-western region of Rajasthan, i.e., a part of the Thar desert. It consists of Rajasthan feeder canal (with the first 167 km in Punjab and Haryana and the remaining 37 km in Rajasthan) and 445 km Rajasthan main canal entirely in Rajasthan.

Rajghat Dam Project (Madhya Pradesh): The Rajghat Dam and Rajghat Hydro Electric Projects are Inter-State projects of MP and UP. The Rajghat Dam is almost complete. All the three units of Rajghat Hydro-Electric Project had been synchronized during 1999 and power generation has been continuing ever since.

Ramganga (Uttarakhand): A dam across Ramganga, a tributary of the Ganga river located in Garhwal district. The project has, besides reducing the intensity of floods in central and western Uttar Pradesh, provided water for the Delhi water supply scheme. Ranjit Sagar Dam (Thein Dam) (Punjab): A multi-purpose highest dam in the country, built on the Ravi river for the benefit of Punjab, Haryana and Jammu and Kashmir.

Rihand Project (Uttar Pradesh and Madhya Pradesh): It is the largest man-made lake in India on the borders of Uttar Pradesh and Madhya Pradesh with a capacity of 300 MW annually.

Sabarmati (Gujarat): A storage dam across Sabarmati river near Dhari Village in Mehsana district and wasna barrage near Ahmedabad.

Salal Project (Jammu & Kashmir): With the successful completion of the 2.5-km long tailrace tunnel, the 690-MW Salal (Stage I and II) project in Jammu and Kashmir became fully operational on August 6, 1996.

Sarda Sahayak (Uttar Pradesh): A barrage across the river Ghaghra, a link channel, a barrage across River Sarda and a feeder channel of two major aqueducts over rivers Gomti and Sai.

Sharavathi Project (Karnataka): It is located at the Jog Falls with a capacity of 891 MW. It primarily feeds Bengaluru industrial region and also Goa and Tamil Nadu.

Sone High Level Canal(Bihar): An extension on Sone Barrage project.

Tawa Project (Madhya Pradesh): A project across the Tawa river, a tributary of the Narmada in Hoshangabad district.

Tehri Dam Project (Uttarakhand): Earth and rock-fill dam on Bhagirathi river in Tehri district.

Tungabhadra Project (Joint Project of Andhra Pradesh and Karnataka): On the Tungabhadra River.

Ukai Project (Gujarat): A multipurpose project across Tapti river near Ukai village.

Upper Krishna Project (Karnataka): A project consisting of Narayanpur dam across the Krishna river and a dam at Almatti.

Upper Penganga Project (Maharashtra): Two reservoirs on Penganga river at Isapur in Yavatmal district and the other on Rayadhu river at Sapli in Parbhani district.

Uri Power Project (Jammu & Kashmir): It is located on the river Jhelum in the Uri Tehsil of Baramulla district in Jammu & Kashmir. It is a 480-MW hydroelectric project which was dedicated to the nation of February 13, 1997.

IMPORTANCE OF IRRIGATION IN INDIA

India is a Tropical country with varieties of climate, topography and vegetation. We have an agrarian economy and many resources depend on the agricultural output. For maximum crop productivity it is very essential to supply the optimum quantity of irrigation water at proper timings. Irrigation water performs number of important function, e.g. acts as nutrient and moisture carrier, ensures better utilization of chemical fertilizer, and maintain optimum temperature around plant environment.

Irrigation is necessary for the following reasons:

- Scanty Rainfall
- Non-Uniform Rainfall
- Increasing Yield in Dry Farming
- To Practice Crop Rotation

- Controlled Water supply
- To wash out or dilutes salts in soil
- To cool the soil and also the atmosphere and makes more favourable environment for healthy plant growth

PROBLEMS WITH EXCESSIVE IRRIGATION

The excess irrigation and unscientific use of irrigation water may give rise to the following ill effects. The states like Punjab are experiencing ill effects of excessive irrigation as part of the state is facing very deep water table problem and another part is facing problem of salinity. Proper management of water is necessary for good natural resource management and sustainable yield. The disadvantages of excessive irrigation may be listed as follows:

- Wasteful use of water
- Water logging
- Soil degradation in irrigated areas
- Contamination of water with harmful substances
- Damp climate and Ecological imbalances
- Mosquitoes breeding.

Types of Irrigation

Sources of irrigation water are either surface water direct or stored or ground water to be tapped through bored wells using pumping system.

Irrigation may broadly be classified into Surface Irrigation and Sub-surface irrigation.

Surface Irrigation

In this type of irrigation water wets the soil surface. It can be further classified into:

Flow Irrigation

When water is supplied from higher level to lower level by the action of gravity then it is called flow irrigation. The irrigation from canal water or river water is the example of flow irrigation.

Flow irrigation can be further subdivided into:

(a) Perennial irrigation

(b) Flood irrigation

Perennial Irrigation

In this type of irrigation system, the water requirement for irrigation is supplied constantly and continuously in accordance with crop requirements throughout the crop period. In this system of irrigation, water is supplied through the storage canal head works and canal distribution system.

Perennial canal system may be further sub-divided as follows:

Direct Irrigation

When irrigation is done from direct run off of a river, or by diverting the river run off water into some canal by constructing a diversion weir or barrage across the river. For example, Ganga Irrigation canal system.

Storage Irrigation

When a dam is constructed across a river to store water during the monsoon and the stored water is supplied in the off taking channels during periods of low flow, it is called storage irrigation. For example, Ram Ganga dam project in UP. In coastal areas where rivers are not perennial storage irrigation becomes a necessity. However, it is more costly and difficult to construct.

Flood Irrigation

This kind of irrigation is sometimes called as Inundation irrigation. In this method of irrigation soil is kept submerged and flooded with water, so as to cause thorough saturation of the land.

Lift Irrigation

When water is lifted up by any manual or mechanical means such as Persian wheel, pumps, etc. and then supplied for irrigation then it is called lift irrigation.

LOCALIZED IRRIGATION

Drip Irrigation

Drip irrigation, also known as trickle irrigation, functions as its name suggests. In this system water falls drop by drop just at the position of roots. Water is delivered at or near the root zone of plants, drop by drop. This method can be the most water-efficient method of irrigation, if managed properly, since evaporation and runoff are minimized.

The drip irrigation system is particularly suited to areas where water quality is marginal, land is steeply sloping or undulating and of poor quality, where water or labour are expensive, or where high value crops require frequent water applications.

It is more economical for orchard crops than for other crops and vegetables since in the orchards plants as well as rows are widely spaced. Drip irrigation limits the water supplied for consumptive use of plants. By maintaining a minimum soil moisture in the root zone, thereby maximizing the water saving. A unique feature of drip irrigation is its excellent adaptability to saline water. Since the frequency of irrigation is quite high, the plant base always remains wet which keeps the salt concentration in the plant zone below the critical. Irrigation efficiency of a drip irrigation system is more than 90 percent. Drip irrigation usage in India is expanding rapidly

In modern agriculture, drip irrigation is often combined with plastic mulch, further reducing evaporation, and is also the means of delivery of fertilizer. The process is known as fertigation.

Fertigation

Fertigation is the application of fertilizers, soil amendments, or other water-soluble products through an irrigation system. In this system fertilizer solution is distributed evenly in irrigation. The availability of nutrients is very high therefore the efficiency is more.

Benefits of fertigation

- Increased nutrient absorption by plants
- Reduction in fertilizer and chemicals needed
- Reduced leaching to the water table
- Reduction in water usage due to the plant's resulting increased root mass's ability to trap and hold water
- Application of nutrients at the precise time they are needed and at the rate they are utilized

Disadvantages of Fertigation

- Concentration of solution decreases as fertilizer dissolves, leading to poor nutrient placement
- Results in pressure loss in main irrigation line
- Limited capacity
- Use of chemical fertilizers of low-sustainability, instead of organic fertilizers.
- Dependent on water supply's non restriction by drought rationing.

SPRINKLER SYSTEM

In sprinkler or overhead irrigation, water is piped to one or more central locations within the field and distributed by overhead high-pressure sprinklers or guns. The pump unit is usually a centrifugal pump which takes water from a source and provides adequate pressure for delivery into the pipe system Sprinkler irrigation has high efficiency. It however, varies according to climatic conditions; 60% in warm climate; 70% in moderate climate and 80% in humid or cool climate. Sprinklers are, however, not suitable for soils which easily form a crust. The water that is pumped through the pump pipe sprinkler system must be free of suspended sediments

Sub-surface Irrigation

In this type of irrigation, water does not wet the soil surface.

In this system of irrigation the supplied water comes directly in touch with root zone of the crops. This system of irrigation may be employed usefully under the following conditions:

- Topography conditions of area are uniform.
- Land slope is moderate.
- The quality of irrigation water is good.
- The soil in the roof zone is permeable in nature.

Sub surface irrigation may be classified into two types

Natural Sub-irrigation

When underground irrigation can be achieved simply by natural processes without any extra efforts it is called natural sub-Irrigation. In fact leakage water from channel, etc. goes underground and during seepage through the sub soil, it may irrigate the crop in the lower lands

Artificial Sub-irrigation

The open jointed system of drain is artificially laid below the soil so as to supply the water to the crop by capillary action, then it is known as artificial sub-irrigation. This process is not adopted in India because it is very costly.

IRRIGATION SCHEMES IN INDIA

Pradhan Mantri Krishi Sinchai Yojana (PMKSY)

The vision of PMKSY is to ensure access to some means of protective irrigation to all agricultural farms in the country – to produce 'per drop more crop. Thus bringing much desired rural prosperity. PMKSY is strategized by focusing on end-to end solution in irrigation supply chain, viz. water sources, distribution network, efficient farm level applications, extension services on new technologies and information etc. based on comprehensive planning process at district/State level.

Irrigation Water Management

Mission Kakatiya

Mission Kakatiya is a flagship program under Telangana government aimed at restoring minor irrigation sources of water like ponds and tanks. The objective is to enhance the development of Minor Irrigation infrastructure, strengthening community based irrigation management in a decentralized manner and to adopt a comprehensive programme for restoration of tanks and sources of water to effectively utilize 265 TM of water allocated for minor irrigation sector under Godavari and Krishna river Basins. Gram Sabha were conducted and the proposed plans were discussed with the villagers. Farmers were motivated to co-operate and were suggested to deposit the silt for field preparation. Several district level coordination committees were formed. Emphasis was given for improvement in delivery time of services. Steps like tank de-siltation, restoration of the feeder channels, re-sectioning of irrigation channels, repair of bund, weir and sluice, raising of FTL (Full Tank Level) wherever required were carried out.

Importance of Irrigation in India

Increase in agricultural production and productivity depends, to a large extent, on the availability of water. Hence, the importance of irrigation is however, the availability of irrigation facilities which is highly inadequate in India. For example, in 1950-51, gross irrigated area as percentage of gross cropped area was only 17%. Even now 60% of gross cropped area depends on rain. That is why Indian agriculture is called a gamble in the monsoon.

1. Control of Drought and Famines Insufficient, uncertain and irregular rain causes uncertainty in agriculture. The period of rain is restricted to only four months in a year, June to September, when monsoon arrives. The remaining eight months are dry. There is some rainfall during the months of December and January in some parts of the country. Even during monsoon , the rainfall is scanty and undependable in many parts of the country. Sometimes the monsoon delayed considerably while sometimes they cease prematurely. This pushes large areas of the country into drought conditions. With the help of irrigation, droughts and famines can be effectively controlled.

2. Higher productivity on irrigated land: Productivity on irrigated land is considerably more than the productivity on un-irrigated land.
3. Multiple cropping possible: Since India has a tropical and sub-tropical climate, it has potentialities to grow crops on a year round basis. However, since 80% of the annual rainfall is received in less than four months, multiple cropping is generally not possible. Provision of irrigation facilities can make possible the growing of two or three crops in a year in most areas of the country. This will considerably enhance agriculture production and productivity.
4. Role in new agricultural strategy: The successful implementation of the High Yielding Programme enhances agricultural production in a great intent.
5. Bringing more land under cultivation: Total reporting area for land utilization statistics was 306.05 million hectares in 1999-2000. of this 19.44 million hectares was current fallow land. Current fallowed include lands which are lying fallow for less than one year other than current fallows includes land lying un-ploughed for one to five years. Cultivable waste land comprises another 13.83 million hectares. Cultivation on all such lands is impossible in some cases while in others it requires substantial capital investment to make land fit for cultivation. Provision of irrigation facilities can make some portion of this land cultivable.
6. Reduces instability in output levels: Irrigation help in stabilising the output and yield levels. It also plays a protective role during drought years. Since, both income and employment are positively and closely related to output, prevention of fall in output during drought is an important instrument for achieving stability of income and employment in the countryside. Irrigation has enabled many states to acquire 'partial immunity' from drought.
7. Indirect benefits of irrigation: Irrigation confers indirect benefits through increased agricultural production. Employment potential of irrigated lands, increase production, helps in developing allied activities means of water transport etc. are improve income of government from agriculture. Availability of regular water supply will increase the income of farmers imparting a sense of security and stability in agriculture.

5

PLANNING OF RESERVOIRS

A dam constructed in a valley relies on the natural topography to provide most of the basin of the reservoir. Dams are typically located at a narrow part of a valley downstream of a natural basin. The valley sides act as natural walls with the dam located at the narrowest practical point to provide strength and the lowest practical cost of construction. In many reservoir construction projects people have to be moved and re-housed, historical artifacts moved or rare environments relocated. Examples include the temples of Abu Simbel (which were moved before the construction of the Aswan Dam to create Lake Nasser from the Nile in Egypt) and the re-location of the village of Capel Celyn during the construction of Llyn Celyn.

Construction of a reservoir in a valley will usually necessitate the diversion of the river during part of the build often through a temporary tunnel or by-pass channel. In hilly regions reservoirs are often constructed by enlarging existing lakes. Sometimes in such reservoirs the new top water level exceeds the watershed height on one or more of the feeder streams such as at Llyn Clywedog in Mid Wales. In such cases additional side dams are required to contain the reservoir. Where the topography is poorly suited to a single large reservoir, a number of smaller reservoirs may be constructed in a chain such as in the River Taff valley where the three reservoirs Llwyn-on Reservoir, Cantref Reservoir and Beacons Reservoir form a chain up the valley.

The main factors to be considered in the planning of a reservoir are:

(i) Rim stability

(ii) Water-holding capability

(iii) Evaporation

(iv) Storage zones

(v) Seismicity

(vi) Sedimentation

Rim stability and water-holding capability are interrelated. Rim failure can be caused due to either sliding or erosion of a segment of the reservoir rim. Seepage of water is mainly responsible for such failures. Major slides into a reservoir would, obviously, reduce the reservoir capacity considerably.

Similarly, snow avalanches and masses of ice falling from hanging glaciers can cause serious problems. Besides reducing the capacity of the reservoir, a rapidly moving slide may also generate waves. A dam may be overtopped due to the resulting wave action or rise of the water surface on account of a major slide into the reservoir.

If the reservoir site is likely to be affected by the slides and it cannot be abandoned, some restraining steps in reservoir operation should be taken to avoid serious failure. These steps could be in the form of limiting the filling and drawdown rates or imposing the maximum allowable water surface at a level lower than the maximum normal water surface.

Alternatively, installation of drains to relieve water pressure along likely slip surfaces, some form of impervious lining and pinning the unstable mass to its parent formation by rock bolting can be resorted to for preventing slides.

Stabilisation of the unstable mass can also be achieved by strengthening or replacing weak material. Grouting is the most common remedy for strengthening such weak masses. It may be desirable to plan the steps to be taken to mitigate the effects of a potential slide after it has occurred in spite of all preventive steps.

Reservoir water loss either to the air or to the ground can be a controlling factor in the selection of site for a conservation reservoir. For a flood control reservoir, water loss is of concern only if it relates to safety of the project.

Lining of the surface through which seepage is expected is one of the preventive measures to reduce the reservoir water loss to the ground. At times, a blanket of impervious material extending from the heel of the dam is required. This too serves to control the seepage from the reservoir.

Loss of reservoir water to the air occurs due to direct evaporation from the reservoir surface. The evaporation losses are affected by the climate of the region, shape of the reservoir, wind conditions, humidity and temperature.

From the considerations of evaporation, a reservoir site having a small surface area to volume ratio will be better than a saucer-shaped reservoir of equal capacity. Evaporation-retardant chemicals increase the surface tension of water by forming mono-molecular film and, thus, reduce evaporation.

The maximum elevation to which water level in a reservoir generally reaches during normal operating conditions is known as full (or maximum) reservoir level. This level usually coincides with the level of the spillway crest.

The minimum elevation to which water level in a reservoir can be drawn under normal operating conditions is called minimum reservoir level (or dead storage level). This level is fixed by the level of the lowest outlet in the dam. The volume of water stored below this level is not available for use under normal conditions and, accordingly, this storage is termed as dead storage.

The storage volume between the minimum and maximum reservoir levels is called useful storage (or live storage). During floods the reservoir level exceeds the full reservoir level and water starts flowing over the spillway. The storage volume between the maximum reservoir level and the reservoir level when the design flood passes over the spillway is called the surcharge storage.

Bank storage is the water which spreads out from a body of water, filling interstices of the surrounding earth and rock mass. This water is assumed to remain in the surrounding mass and does not continue to move to ultimately join the ground water or surface water as seepage water does.

The storage zone is theoretically recoverable but, in a sense, is considered a loss. Storage zone is not mitigable. It must, however, be estimated for feasibility investigations and measured during reservoir operation for providing guidelines for reservoir regulation.

It appears that there is some effect of reservoir impoundment on the increased seismic activity of an area in which a large reservoir (having a storage capacity of more than 12 x 108 m3 behind a dam higher than 90 m) has been constructed. However, there have been large reservoirs without increasing the seismic activity of the region.

The increased seismic activity is attributed to the changes in the normal effective stresses in the underlying rock because of the increased pore pressure. The transmission of hydrostatic pressure through discontinuities in the underlying rock can have a triggering effect where a critical state of stress already exists. The relationship between the reservoir impoundment and the earthquake relationship is not fully understood.

Hence, it is necessary that every large reservoir site is subjected to detailed geologic, geodetic and seismic studies for feasibility decision. These observations must be continued during the reservoir operation too in order to better understand the relationship between reservoir impoundment and seismic activity.

The streams bringing water to the reservoir bring sediments too. The sediment gets deposited in the reservoir due to the reduced stream velocity. The capacity of the reservoir is reduced on account of sediment deposition in the reservoir. Usually, a portion of the reservoir storage is reserved for the storage of the sediment.

The life of a reservoir is predicted on the basis of the amount of sediment delivered to it, the reservoir size and its ability to retain the sediment. Sediment deposition at the initial stage may be beneficial in the sense that it may have the effect of a natural blanket resulting in reduced seepage loss.

Measures to minimise sediment deposition in reservoirs include watershed protection through a vegetative management programme to prevent soil erosion, silt detention basins at inlets of smaller reservoirs and low level outlets in dams to provide flushing action for removal of sediment from the reservoir.

Of the various measures, the watershed protection is the most effective as well as the costliest method. Erosion in the catchment area can be reduced by soil conservation methods, such as afforestation, control of deforestation, control of grazing, checking gully formation by providing small embankments, etc.

SELECTION OF SITE FOR A RESERVOIR:

Based on planning and other considerations, the guidelines for selection of site for a reservoir are as follows:

(i) Availability of a suitable site for construction of dam.

(ii) The hills surrounding the reservoir and the bed of the reservoir should be impervious.

(iii) Availability of good storage capacity with minimum submergence of the adjacent land.

(iv) A reservoir should not be sited downstream of such tributaries which bring-in excess sediment into the river.

(v) Availability of deep gorge which results in larger capacity with lesser water surface area and, therefore, lesser evaporation loss.

(vi) The cost of other associated works is less.

(vii) The site with the possibility of land slides into the reservoir must be avoided.

(viii) The site should not be, as far as possible, on valuable land being used for some other purposes, such as agriculture, forestry, communication and habitation by people,

(ix) Sites with mineral deposits in and around the reservoir area should also be avoided.

Classification of Reservoirs

Depending upon the purposes served, the reservoirs may be classified into following categories.

- Storage or conservation reservoirs
- Flood control reservoirs
- Distribution reservoirs
- Multipurpose reservoirs.

1. Storage or Conservation Reservoirs.

These reservoirs are primarily used to maintain minimum supplies of water for irrigation, hydroelectric generation, domestic and industrial water supply schemes, etc.

During lean months of discharge in the rivers.

We know that discharge in the rivers remains changing day to day, and season to season.

During high floods the excess water in the river goes waste while in dry months it may not be sufficient to meet the minimum needs.

The storage reservoir is constructed to store the excess water of floods and released gradually as and when required.

2. Flood Control Reservoirs.

This reservoir is also called flood mitigation reservoir.

The main purpose of this reservoir is to temporarily store the flood water and release slowly at a safe rate after the floods, so that it may not cause any damage on the down stream side.

So this reservoir may be said a flood prevention reservoir also.

This reservoir requires provision of large spillways and sluice-ways so that excess stored flood water is rapidly releas downstream, but only at the safe rate. Figure 9.1 shows hydrographs at certain point on the river.

ABC is the natural hydrograph at the dam site having a maximum flood discharge

After construction of the dam the hydrograph gets modified because of the development of a reservoir at the back of the dam.

AB'DC is the modifie hydrograph showing maximum flood discharge as Q2.

Thus, because of reservoir the flood discharge has reduce from Q1 to Q2.

Shaded are ABDB'A represents the storage to be provid in the reservoir.

The area DEC represents the excess volume being released from the reservoir after the floods have receded.

Flood control reservoirs may be further classified into two categories.

(i) Storage reservoir of Detention basins.

(ii) Retarding basins or retarding reservoirs.

(i) Storage reservoirs or Detention basins. The reservoirs whose spillways and sluice outlets are fitted with gates and valves, are known as storage reservoirs or detention basins. Gated spillways and gated sluiceways provide more flexibility in operation. They help in exercising better control on the reservoir and thus reservoir water can be used more wisely and usefully.

This reservoir is costly as it involves cost of gates and valves.

Original hydrograph ABC modified hydrograph ABC.

(ii) Retarding basins or retarding reservoirs. In these reservoirs there are no gates at spillways and sluice outlets. In this case sluice and spillway's joint maximum discharging capacity is at the most equal to the maximum safe carrying capacity of the channel down stream.

As floods occur, the reservoir first of all gets filled up, upto normal level.

At this time sluice outlets are discharging out water from the reservoir.

As level in the reservoir increases further, discharge through sluice ways also increases.

At certain level of reservoir, water also starts escaping through ungated spillways.

As the level of water further rises the discharge over spill ways also increases.

At some particular level a balance will be struck between the inflow and outflow of the reservoir.

At this time level of water in the reservoir will become stable.

This condition happens only when flood inflow in the reservoir is equal to the outflow from the reservoir.

Now when floods occurring recede, inflow in the reservoir will also decrease but the outflow is at the same maximum rate.

Hence flood water which had accumulated in the reservoir will now be slowly flowing out of the reservoir.

This reservoir has the following advantages over detention basin.

(a) Gates are not required to be provided at sluice ways and spillway crests.

(b) Since there are no gates at the spillways, chances of human error in opening the gates during floods cannot take place.

(c) Since water from the reservoir is driven out in few days after floods, the land during maximum floods remains submerged only temporarily.

This submerged land can be used for growing very good crops.

But no habitation on this land should be allowed.

Retarding basins are preferred on small rivers.

3. Distribution Reservoir.

It is a small capacity reservoir which is mainly constructed to meet the water supply requirements of a particular city.

It is made of masonry or cement concrete and may be covered from the top.

This reservoir is filled by treated water at some constant rate.

Since demand of water remains fluctuating during the day, water may have to be drawn from this reservoir at times at rate much more than the inflow rate.

Hence these reservoirs allow pumping units and treatment units to work at predetermined constant rates.

During no demand or very little demand the water coming from treatment units and pump units goes on storing in the reservoir.

During peak demand this stored water from the reservoir is used to make up the required supplies.

4. Multipurpose Reservoirs.

The reservoir planned and designed keeping only one purpose in view is known as single-purpose reservoir.

Reservoir planned and designed keeping more than one purpose in view is called Multi purpose reservoir.

For example a reservoir designed to protect the downstream areas from floods and also to conserve water for irrigation, water-supply, hydroelectric purpose etc. shall be called a Multi-purpose reservoir.

Bhakra Dam, Nagarjun Sagar Dam are the examples of important multi-purpose projects of India.

Bank-side reservoir

Where water is taken from a river of variable quality or quantity, bank-side reservoirs may be constructed to store the water pumped or siphoned from the river. Such reservoirs are usually built partly by excavation and partly by the construction of a complete encircling bund or embankment which may exceed 6 km in circumference.

Both the floor of the reservoir and the bund must have an impermeable lining or core, often made of puddled clay. The water stored in such reservoirs may have a residence time of several months during which time normal biological processes are able to substantially reduce many contaminants and almost eliminate any turbidity.

The use of bank-side reservoirs also allows a water abstraction to be closed down for extended period at times when the river is unacceptably polluted or when flow conditions are very low due to drought. The London water supply system is one example of the use of bank-side storage for all the water taken from the River Thames and River Lee with many large reservoirs such as Queen Mary Reservoir visible along the approach to London Heathrow Airport.

Service reservoir

Service reservoirs store fully treated potable water close to the point of distribution. Many service reservoirs are constructed as water towers, often as elevated structures on concrete pillars where the landscape is relatively flat. Other service reservoirs are entirely underground, especially in more hilly or mountainous country. In the United Kingdom, Thames Water has many underground reservoirs in London built in the 1800s by the Victorians, most of which are lined with brick. Honor Oak Reservoir, which was completed in 1909, is believed to one of the largest of this type in Europe. The roof is supported on large brick pillars and arches and the outside surface is grassed over.

Service reservoirs perform several functions including ensuring sufficient head of water in the water distribution system and providing hydraulic capacitance in the system to even out peak demand from consumers enabling the treatment plant to run at optimum efficiency. Large service reservoirs can also be managed to so that energy costs in pumping are reduced by concentrating refilling activity at times of day when power costs are low.

Five thousand years ago, the craters of extinct volcanoes in Arabia were used as reservoirs by farmers for their irrigation water. Dry climate and water scarcity in India led to early development of water management techniques, including the building of a reservoir at Girnar in 3000 BC. Artificial lakes dating to the 5th century BC have been found in ancient Greece. An artificial lake in present-day Madhya Pradesh province of India, constructed in the 11th century, covered 650 square metres (7,000 sq ft).

In Sri Lanka large reservoirs have been created by ancient Sinhalese kings in order to save the water for irrigation. The famous Sri Lankan king Par?kramab?hu I of Sri Lanka stated " do not let a drop of water seep into the ocean without benefiting mankind ". He created the reservoir named Parakrama Samudra(sea of King Parakrama), which has astonished archaeologists.

Uses

Direct water supply

Many dammed river reservoirs and most bank-side reservoirs are used to provide the raw water feed to a water treatment plant which delivers drinking water through water mains. The reservoir does not simply hold water until it is needed; it can also be the first part of the water treatment process. The time the water is held for before it is released is known as the retention time. This is a design feature that allows particles and silts to settle out, as well as time for natural biological treatment using algae, bacteria and zooplankton that naturally live within the water.

However natural limnological processes in temperate climate lakes produces temperature stratification in the water body which tends to partition some elements such as manganese and phosphorus into deep, cold anoxic water during the summer months. In the autumn and winter the lake becomes fully mixed again. During drought conditions, it is sometimes necessary to draw down the cold bottom water and the elevated levels of manganese in particular can cause problems in water treatment plants.

Hydroelectricity

A reservoir generating hydroelectric includes turbines connected to the retained water body by large diameter pipes. These generating sets may be at the base of the dam or some distance away. Some reservoirs generating hydro-electricity use pumped re-charge in which a high level reservoir is filled with water using high performance electric pumps at times when electricity demand is low and then uses this stored water to generate electricity by releasing the stored water into a low level reservoir when electricity demand is high. Such systems are called pump storage schemes.

Controlling watercourses

Reservoirs can be used in a number of ways to control how water flows through downstream waterways.

Downstream water supply - water may be released from an upland reservoir so that it can be abstracted for drinking water lower down the system, sometimes hundred of miles further down downstream

Irrigation - water in an irrigation reservoir may be released into networks of canals for use in farmlands or secondary water systems. Irrigation may also be supported by reservoirs which maintain river flows allowing water to be abstracted for irrigation lower down the river.

Flood control - also known as an "attenuation" or "balancing" reservoir, flood control reservoirs collect water at times of very high rainfall, then release it slowly over the course of the following weeks or months. Some of these reservoirs are constructed across the river line with the onward flow controlled by an orifice plate. When river flow exceeds the capacity of the orifice plate water builds behind the dam but as soon as the flow rate reduces the water behind the dam slowly releases until the reservoir is empty again. In some cases such reservoirs only function a few times in a decade and the land behind the reservoir may be developed as community or recreational land. A new generation of balancing dams are being developed to combat the climatic consequences of climate change. They are called "Flood Detention Reservoirs". Because these reservoirs will remain dry for long periods, there may be a risk of the clay core drying out reducing its structural stability. Recent developments include the use of composite core fill made from recycled materials as an alternative to clay.

Canals - Where a natural watercourse's water is not available to be diverted into a canal, a reservoir may be built to guarantee the water level in the canal; for example, where a canal climbs to cross a range of hills through locks.

Recreation - water may be released from a reservoir to artificially create or supplement white-water conditions for kayaking and other white-water sports. On salmonid rivers special releases (in Britain called freshets) are made to encourage natural migration behaviours in fish and to provide a variety of fishing conditions for anglers.

Flow balancing

Occasionally reservoirs are used to balance the flow in highly managed systems, taking in water during high flows and releasing it again during low flows. In order for this to work without pumping requires careful control of water levels using adjustable sluices. One example of this is Llyn Tegid in North Wales. This is a natural lake whose level was raised by a low dam and into which the River Dee flows or discharges depending upon flow conditions at the time as part of the River Dee regulation system. This mode of operation is a form of hydraulic capacitance in the river system.

Recreation

The water bodies provided by many reservoirs often allow some recreational uses such as fishing, boating, and other activities. Special rules may apply for the safety of the public and to protect the quality of the water and the ecology of the surrounding area. Many reservoirs now support and encourage less informal and less structured recreation such as natural history, bird watching, landscape painting, walking and hiking and often provide information boards and interpretation material to encourage responsible use.

Operation

Water falling as rain upstream of the reservoir together with any groundwater emerging as springs is stored in the reservoir. Any excess water can be spilled via a specifically designed spillway. Stored water may be piped by gravity for use as drinking water, to generate hydro-electricity or to maintain river flows to support downstream uses. Occasionally reservoirs can be managed to retain high rain-fall events to prevent or reduce downstream flooding. Some reservoirs support several uses and the operating rules may be complex.

Most modern reservoirs have a specially designed draw-off tower that can discharge water from the reservoir at different levels both to access water as the reservoir draws down but also to allow water of a specific quality to be discharged into the downstream river as compensation water.

The operators of many upland or in-river reservoirs have obligations to release water into the downstream river to maintain river quality, support fisheries, maintain downstream industrial uses. maintain recreational use or for a range of other requirements. Such releases are known as compensation water.

Terminology

The terminology for reservoirs varies from country to country. In most of the world reservoir areas are expressed in km^2 whilst in the USA acres are commonly used. For volume either m^3 or km^3 are widely used with acre feet used in the USA.

The capacity, volume or storage of a reservoir is usually divided into distinguishable areas. Dead or inactive storage refers to water in a reservoir that cannot be drained by gravity through a dam's outlet works, spillway or power plant intake and can only be pumped out. Dead storage allows sediments to settle which improves water quality and also creates hydraulic head along with an area for fish during low levels. Active or live storage is the portion of the reservoir that can be utilized for flood control, power production, navigation and downstream releases.

In addition, a reservoir's flood control capacity is the amount of water it can regulate during flooding. The surcharge capacity is the capacity of the reservoir above the spillway crest that cannot be regulated. In the United States the water below the normal maximum level of a reservoir is called the conservation pool. In the UK top water level describes the reservoir full state whist fully drawn down describes the minimum retained volume.

Modelling reservoir management

There is a wide variety of software for modelling reservoirs, from the specialist Dam Safety Program Management Tools (DSPMT) to the relatively simple WAFLEX, to integrated models like the Water Evaluation And Planning system (WEAP) that place reservoir operations in the context of system-wide demands and supplies.

Safety

In many countries large reservoirs are closely regulated to try and prevent or minimise failures of containment. Whilst much of the effort is directed at the dam and its associated structures as the weakest part of the overall structure, the aim of such controls is to prevent an uncontrolled release of water from the reservoir. Reservoir failures can generate huge increases in flow down a river valley with the potential to wash away towns and villages and cause considerable loss of life such as the devastation following the failure of containment at Llyn Eigiau which killed 17 people.

A notable case of reservoirs being used as an instrument of War involved the British Royal Air Force Dambusters raid on Germany in World War II (codenamed "Operation Chastise"), in which three German reservoir dams were selected to be breached in order to impact on German infrastructure and manufacturing and power capabilities deriving from the Ruhr and Eder rivers. The economic and social impact was derived from the enormous volumes of previously stored water that swept down the valleys wreaking destruction. This raid later became the basis for several films.

All reservoirs will have a monetary cost/benefit assessment made before construction to see if the project is worth proceeding with. However, such analysis can often omit the environmental impacts of dams and the reservoirs that they contain. Some impacts such as the greenhouse gas production associated with concrete manufacture are relatively easy to estimate. Other impact on the natural environment and social and cultural effects can be more difficult to assess and to weigh in the balance but identification and quantification of these issues are now commonly required in major construction projects in the developed world.

Climate Change

Depending upon the circumstances, a reservoir built for hydro-electricity generation can either reduce or increase the net production of greenhouse gases. An increase can occur if plant material in the flooded areas decays in an anaerobic environment releasing (methane and carbon dioxide). This apparently counter intuitive position arises because much carbon is released as methane which is approximately 8 times more potent as a greenhouse gas than carbon dioxide.

A study for the National Institute for Research in the Amazon found that Hydroelectric reservoirs release a large pulse of carbon dioxide from above-water decay of trees left standing in the reservoirs, especially during the first decade after closing. This elevates the global warming impact of the dams to levels much higher than would occur by generating the same power from fossil fuels.

According to the World Commission on Dams report (Dams And Development), when the reservoir is relatively large and no prior clearing of forest in the flooded area was undertaken, greenhouse gas emissions from the reservoir could be higher than those of a conventional oil-fired thermal generation plant. For instance, In 1990, the impoundment behind the Balbina Dam in Brazil (inaugurated in 1987) had over 20 times the impact on global warming than would generating the same power from fossil fuels, due to the large area flooded per unit of electricity generated.

A decrease can occur if the dam is used in place of traditional power generation, since electricity produced from hydroelectric generation does not give rise to any flue gas emissions from fossil fuel combustion (including sulfur dioxide, nitric oxide and carbon monoxide from coal). The Tucurui dam in Brazil (closed in 1984) had only 0.4 times the impact on global warming than would generating the same power from fossil fuels. Dams can produce a block for migrating fish, trapping them in one area, producing food and a habitat for various water-birds. They can also flood various ecosystems on land and may cause extinctions.

Human Impact

Dams can severely reduce the amount of water from reaching countries downstream of them, causing water stress between the countries, e.g. the Sudan and Egypt, which damages farming businesses in the downstream countries, and reduces drinking water.

Farms and villages, e.g. Ashopton can be flooded by the creation of reservoirs, ruining many livelihoods. For this very reason, worldwide 80 million people have had to be forcibly relocated due to dam construction.

Limnology

The limnology of reservoirs has many similarities to that of lakes of equivalent size. There are however significant differences. Many reservoirs experience considerable variations in level producing significant areas that are intermittently underwater or dried out. This greatly limits the productivity or the water margins and limits the number of species able to survive in these conditions.

Upland reservoirs tend to have a much shorter residence time than natural lakes and this can lead to more rapid cycling of nutrients through the water body so that they are more quickly lost to the system. This may be seen as a mismatch between water chemistry and water biology with a tendency for the biological component to be more oligotrophic than the chemistry would suggest.

Conversely, lowland reservoirs drawing water from nutrient rich rivers, may show exaggerated eutrophic characteristics because the residence time in the reservoir is much greater than in the river and the biological systems have a much greater opportunity to utilise the available nutrients.

Deep reservoirs with multiple level draw off towers can discharge deep cold water into the downstream river greatly reducing the size of any hypolimnion. This in turn can reduce the concentrations of phosphorus released during any annual mixing event and may therefore reduce productivity.

The Dams in front of reservoirs act as knickpoints-the energy of the water falling from them reduces and deposition is a result below the Dams.

Seismicity

The filling (impounding) of reservoirs has often been attributed to reservoir-triggered seismicity (RTS) as seismic events have occurred near large dams or within their reservoirs in the past. These events may have been triggered by the filling or operation of the reservoir and are on a small scale when compared to the amount of reservoirs worldwide. Of over 100 recorded events, early examples include the 60 m (197 ft) tall Marathon Dam in Greece (1929), the 221 m (725 ft) tall Hoover Dam in the U.S. (1935).

Most events involve large dams and small amounts of seismicity. The only four recorded events above a 6.0- magnitude (Mw) are the 103 m (338 ft) tall Koyna Dam in India which registered a Mw of 6.3 along with the 120 m (394 ft) Kremasta Dam in Greece which registered a 6.3-Mw as well. Following those two, the next largest were the 122 m (400 ft) high Kariba Dam in Zambia at 6.25-Mw and the 105 m (344 ft) Xinfengjiang Dam in China at 6.1-Mw.

Disputes occur over when RTS has occurred due to a lack of hydrogeological knowledge at the time of the event. It is accepted though that the infiltration of water into pores and the weight of the reservoir do contribute to RTS patterns. For RTS to occur, there must be a seismic structure near the dam or its reservoir and the seismic structure must be close to failure. Additionally, water must be able to infiltrate the deep rock stratum as the weight of a 100 m (328 ft) deep reservoir will have little impact when compared the deadweight of rock on a crustal stress field which may be located at a depth of 10 km (6 mi) or more.

Micro climate

Reservoirs may change the local micro-climate increasing humidity and reducing extremes of temperature. Such effects are claimed by some South Australian winerys as increasing the quality of the wine production.

Sedimentation

Sedimentation is the tendency for particles in suspension to settle out of the fluid in which they are entrained, and come to rest against a barrier. This is due to their motion through the fluid in response to the forces acting on them: these forces can be due to gravity, centrifugal acceleration or electromagnetism.

In geology sedimentation is often used as the polar opposite of erosion, i.e., the terminal end of sediment transport. In that sense it includes the termination of transport by saltation or true bedload transport. Settling is the falling of suspended particles through the liquid, whereas sedimentation is the termination of the settling process.

Sedimentation may pertain to objects of various sizes, ranging from large rocks in flowing water to suspensions of dust and pollen particles to cellular suspensions to solutions of single molecules such as proteins and peptides. Even small molecules such as aspirin can be sedimented, although it can be difficult to apply a sufficiently strong force to produce significant sedimentation.

The term is typically used in geology, to describe the deposition of sediment which results in the formation of sedimentary rock, and in various chemical and environmental fields to describe the motions of often-smaller particles and molecules. Process is also used in biotech industry to separate out cells from the culture media.

In a sedimentation experiment called tripothsis, the applied force accelerates the particles to a terminal velocity vterm at which the applied force is exactly canceled by an opposing drag force. For small enough particles (low Reynolds number), the drag force varies linearly with the terminal velocity, i.e., Fdrag = fvterm (Stokes flow) where f depends only on the properties of the particle and the surrounding fluid. Similarly, the applied force generally varies linearly with some coupling constant (denoted here as q) that depends only on the properties of the particle, Fapp = qEapp. Hence, it is generally possible to define a sedimentation coefficient that depends only on the properties of the particle and the surrounding fluid. Thus, measuring s can reveal underlying properties of the particle.

In many cases, the motion of the particles is blocked by a hard boundary; the resulting accumulation of particles at the boundary is called a sediment. The concentration of particles at the boundary is opposed by the diffusion of the particles.

The sedimentation of particles under gravity is described by the Mason-Weaver equation, which has a simple exact solution. The sedimentation coefficients in this case equals mb / f, where mb is the buoyant mass.

The sedimentation of particles under the centrifugal force is described by the Lamm equation, which likewise has an exact solution. The sedimentation coefficient s also equals mb / f, where mb is the buoyant mass. However, the Lamm equation differs from the Mason-Weaver equation because the centrifugal force depends on radius from the origin of rotation, whereas gravity is presumed constant. The Lamm equation also has extra terms, since it pertains to sector-shaped cells, whereas the Mason-Weaver equation pertains to box-shaped cells (i.e., cells whose walls are aligned with the three Cartesian axes).

Particles with a charge or dipole moment can be sedimented by an electric field or electric field gradient, respectively. These processes are called electrophoresis and dielectrophoresis, respectively. For electrophoresis, the sedimentation coefficient corresponds to the particle charge divided by its drag (the electrophoretic mobility). Similarly, for dielectrophoresis, the sedimentation coefficient equals the particle's electric dipole moment divided by its drag.

In geology, sedimentation is the deposition of particles carried by a fluid flow. For suspended load, this can be expressed mathematically by the Exner equation, and results in the formation of depositional landforms and the rocks that constitute sedimentary record. An undesired increased transport and sedimentation of suspended material is called siltation, and it is a major source of pollution in waterways in some parts of the world. Climate change also affect siltation rates.

WATER QUALITY

Water quality is the physical, chemical and biological characteristics of water. It is a measure of the condition of water relative to the requirements of one or more biotic species and or to any human need or purpose. It is most frequently used by reference to a set of standards against which compliance can be assessed. The most common standards used to assess water quality relate to drinking water, safety of human contact and for the health of ecosystems.

In the setting of standards, agencies make political and technical/scientific decisions about how the water will be used. In the case of natural water bodies, they also make some reasonable estimate of pristine conditions. Different uses raise different concerns and therefore different standards are considered. Natural water bodies will vary in response to environmental conditions. Environmental scientists work to understand how these systems function which in turn helps to identify the sources and fates of contaminants. Environmental lawyers and policy makers work to define legislation that ensure that water is maintained at an appropriate quality for its identified use. The vast majority of surface water on the planet is neither potable nor toxic. This remains true even if sea water in the oceans (which is too salty to drink) is not counted. Another general perception of water quality is that of a simple property that tells whether water is polluted or not. In fact, water quality is a very complex subject, in part because water is a complex medium intrinsically tied to the ecology of the Earth. Industrial pollution is a major cause of water pollution, as well as runoff from agricultural areas, urban stormwater runoff and discharge of treated and untreated sewage (especially in developing countries).

Categories

The parameters for water quality are determined by the intended use. Work in the area of water quality tends to be focused on water that is treated for human consumption or in the environment.

Human consumption

Contaminants that may be in untreated water include microorganisms such as viruses and bacteria; inorganic contaminants such as salts and metals; organic chemical contaminants from industrial processes and petroleum use; pesticides and herbicides; and radioactive contaminants. Water quality depends on the local geology and ecosystem, as well as human uses such as sewage dispersion, industrial pollution, use of water bodies as a heat sink, and overuse (which may lower the level of the water).

In the United States, the U.S. Environmental Protection Agency (EPA) limits the amounts of certain contaminants in tap water provided by public water systems. The Safe Drinking Water Act authorizes EPA to issue two types of standards: primary standards regulate substances that potentially affect human health, and secondary standards prescribe aesthetic qualities, those that affect taste, odor, or appearance.

The U.S. Food and Drug Administration (FDA) regulations establish limits for contaminants in bottled water that must provide the same protection for public health. Drinking water, including bottled water, may reasonably be expected to contain at least small amounts of some contaminants. The presence of these contaminants does not necessarily indicate that the water poses a health risk.

Some people use water purification technology to remove contaminants from the municipal water supply they get in their homes, or from local pumps or bodies of water. For people who get water from a local stream, lake, or aquifer (well), their drinking water is not filtered by the local government.

Environmental water quality

Environmental water quality, also called ambient water quality, relates to water bodies such as lakes, rivers, and oceans. Water quality standards vary significantly due to different environmental conditions, ecosystems, and intended human uses.

Toxic substances and high populations of certain microorganisms can present a health hazard for non-drinking purposes such as irrigation, swimming, fishing, rafting, boating, and industrial uses. These conditions may also affect wildlife which use the water for drinking or as a habitat. Modern water quality laws general specify protection of fisheries and recreational use and require as a minimum,retention of current quality standards.

There is some desire among the public to return water bodies to pristine, or pre-industrial conditions. Most current environmental laws focus of the designation of uses. In some countries these allow for some water contamination as long as the particular type of contamination is not harmful to the designated uses. Given the landscape changes in the watersheds of many freshwater bodies, returning to pristine conditions would be a significant challenge. In these cases, environmental scientists focus on achieving goals for maintaining healthy eco-systems and may concentrate on the protection of populations of endangered species and protecting human health.

Measurement

The complexity of water quality as a subject is reflected in the many types of measurements of water quality indicators. Some of the simple measurements listed below can be made on-site - temperature, pH, dissolved oxygen, conductivity, Oxygen Reduction potential (ORP), turbidity, Secchi Disk depth- in direct contact with the water source in question. More complex measurements that must be made in a lab setting require a water sample to be collected, preserved, and analyzed at another location.

Making these complex measurements can be expensive. Because direct measurements of water quality can be expensive, ongoing monitoring programs are typically conducted by government agencies. However, there are local volunteer programs and resources available for some general assessment. Tools available to the general public are on-site test kits commonly used for home fish tanks and biological assessments.

The following is a list of indicators often measured by situational category:

Drinking water

Filtering water

- Alkalinity
- Color of water
- pH
- Taste and odor (geosmin, 2-methylisoborneol (MIB), etc)
- Dissolved metals and salts (sodium, chloride, potassium, calcium, manganese, magnesium)

- Microorganisms such as fecal coliform bacteria (Escherichia coli), Cryptosporidium, and Giardia lamblia
- Dissolved metals and metalloids (lead, mercury, arsenic, etc.)
- Dissolved organics: colored dissolved organic matter (CDOM), dissolved organic carbon (DOC)
- Radon
- Heavy metals
- Pharmaceuticals
- Hormone analogs

Environmental

Chemical assessment

- (also see salinity)
- Dissolved Oxygen (DO)
- Nitrate-N
- Orthophosphates
- Chemical oxygen demand (COD)
- Biochemical oxygen demand (BOD)
- Pesticides

Physical assessment

- pH
- Temperature
- Total suspended solids (TSS)
- Turbidity

Biological assessment

Biological monitoring metrics have been developed in many places, and one widely used measure is the presence and abundance of members of the insect orders Ephemeroptera, Plecoptera and Trichoptera. (Common names are, respectively, Mayfly, Stonefly and Caddisfly.) EPT indexes will naturally vary from region to region, but generally, within a region, the greater the number of taxa from these orders, the better the water quality. EPA and other organizations in the

United States offer guidance on developing a monitoring program and identifying members of these and other aquatic insect orders.

Individuals interested in monitoring water quality who cannot afford or manage lab scale analysis can also use biological indicators to get a general reading of water quality. One example is the IOWATER volunteer water monitoring program, which includes a benthic macroinvertebrate indicator key.

Flood Storage

The components of online storage works normally include:

(a) An impounding structure - generally comprises an earth or concrete structure across the river and floodplain, behind which the water is stored;

(b) A flow control structure - normally located within the impounding structure to control the outflow from the storage area;

(c) A spillway - to pass floods that are more extreme than those that the reservoir is designed to attenuate.

The flow control structure can be a fixed throttle (such as an orifice, pipe or flume) sized to have little effect on normal flows, but requiring a significant rise in upstream water level to pass larger flows. Sometimes, the control structure incorporates gates, which are normally left open, but are operated during floods to ensure that downstream flows do not overtop the downstream flood defences.

As noted above, an online flood storage reservoir must also be designed to deal safely with more extreme floods than it is designed to attenuate; these could otherwise lead to the water level upstream of the impounding structure rising above safe levels or unacceptable damage to the impounding structure or surrounding land. This normally involves the provision of an overflow weir or spillway to convey the extreme floods. Sometimes the entire impounding embankment - or a major part of it - provides the overflow route for floods, having been designed to withstand the flow velocities involved. In other cases, the flow control structure incorporates the necessary arrangements for passing flood flows, either through gates or over fixed weirs and other structures.

Offline flood storage works generally consist of:

(a) An intake structure to divert water to the storage area when the river flow or level exceeds a pre-determined value;

(b) A storage area that comprises a reservoir separated from the river, formed either by low ground levels (natural or excavated) or by retaining structures (embankments, walls, or a combination of the two);

(c) An outlet structure that returns water from the storage area to the river after the flood peak has passed;

(d) A spillway to pass floods that are more extreme than those that the reservoir is designed to attenuate.

As illustrated in Figure, the intake structure (which is of course an 'offtake' from the perspective of the river) often includes two components that may be located separately:

- A device designed to control flows passed downriver;
- An inflow structure to the storage area.

The device controlling downstream flow can take most of the forms used for the control of online flood storage works. Rather than building a new control structure, a 'natural' control may be used on the river in some cases, or a pre-existing control such as an existing weir in others. In these cases, the offline storage works are designed to make use of the existing stage/discharge characteristics of the river.

Weirs (Rickard et al, 2003) are often favoured for the inflow structure, sometimes in the form of a sideweir (May et al, 2003) along the riverbank. They have the advantage of beginning to operate whenever river levels rise above a given value, but need to be long to limit the rise in river level needed to pass greater flows into the storage area. For this reason, a gated arrangement is often used.

The offline flood storage area generally lies within the floodplain, but isolated by purpose-built walls or embankments. The volume available for storage in the reservoir depends on the water depth that can be obtained, which is controlled by existing ground levels, and the peak flood level that can be accommodated. This depth is often limited, making it necessary for the reservoir to cover a large area. Choosing a site where the ground is low - either naturally or as a result of excavations (for example, for gravel pits) - increases the depth available, but may mean that pumps are needed to empty the reservoir after the flood has passed.

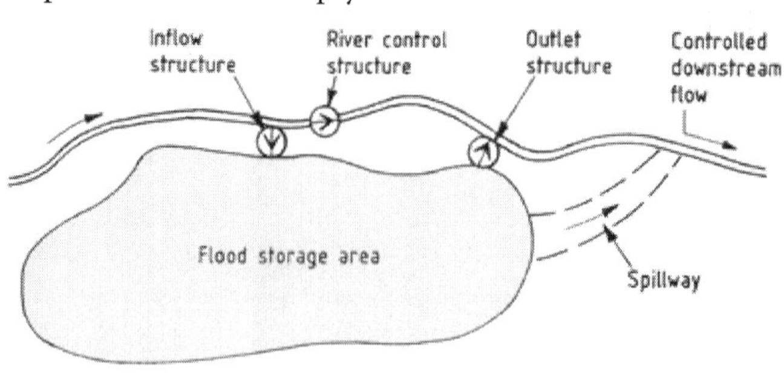

Figure: Offline flood storage area

The outlet may be by gravity (generally using gates), pumped or by a combination of the two (with gravity flow initially, then pumping to drain the lowest sections). The outlet capacity required depends on the volume stored and the time allowed for the system to be fully drained. Whenever the reservoir contains water - and so has less capacity to accommodate another flood - the standard of protection provided will be temporarily compromised. If the outfall is to a tidal river, it may only be possible to complete the emptying the reservoir at low tide, perhaps via a flapgate.

As for flood storage works, arrangements for the safe passage of extreme floods are again needed to protect the reservoir and its associated structures. These may form a major part of the facility, particularly if the inlet to the storage area is uncontrolled.

SEEPAGE: STEADY STATE FLOW OF WATER

If fluid pressures in a soil deposit are uniformly increasing with depth according to $u = p_w g z_w$ then hydrostatic conditions will prevail and the fluids will not be flowing through the soil. zw is the depth below the water table. However, if the water table is sloping or there is a perched water table as indicated in the accompanying sketch, then seepage will occur. For steady state seepage, the seepage velocities are not varying with time. If the water tables are changing levels with time, or if the soil is in the process of consolidation, then steady state conditions do not apply.

Darcy's law

Darcy's law states that the volume of flow of the pore fluid through a porous medium per unit time is proportional to the rate of change of excess fluid pressure with distance. The constant of proportionality includes the viscosity of the fluid and the intrinsic permeability of the soil. For the simple case of a horizontal tube filled with soil

Diagram showing definitions and directions for Darcy's law.

The total discharge, Q (units of volume per time, e.g., ft^3/s or m^3/s) is proportional to the intrinsic permeability, K, the cross sectional area, A, and rate of pore pressure change with distance, , and inversely proportional to the dynamic viscosity of the fluid, ?. The negative sign is needed because fluids flow from high pressure to low pressure. So if the change in pressure is negative (in the x-direction) then the flow will be positive (in the x-direction). The above equation works well for a horizontal tube, but if the tube was inclined so that point b was a different elevation than point a, the equation would not work. The effect of elevation is accounted for by replacing the pore pressure by excess pore pressure, ue defined as:

$$ue = u, p_w gz$$

where z is the depth measured from an arbitrary elevation reference (datum). Replacing u by ue we obtain a more general equation for flow:

Dividing both sides of the equation by A, and expressing the rate of change of excess pore pressure as a derivative, we obtain a more general equation for the apparent velocity in the x-direction:

where vx has units of velocity and is called the Darcy velocity, or discharge velocity. The seepage velocity vsx = (average velocity of fluid molecules in the pores) is related to the Darcy velocity, and the porosity, n

Civil Engineers predominantly work on problems that involve water and predominantly work on problems on earth (in earths gravity). For this class of problems, civil engineers will often write Darcy's law in a much simpler form :

$$v = ki$$

where k is called permeability, and is defined as , and i is called the hydraulic gradient. The hydraulic gradient is the rate of change of total head with distance. The total head, h at a point is defined as the height (measured relative to the datum) to which water would raise in a piezometer at that point. The total head is related to the excess water pressure by:

$$ue = p_w gh + \text{Constant}$$

and the Constant is zero if the datum for head measurement is chosen at the same elevation as the origin for the depth, z used to calculate ue.

Typical values of permeability

Values of the permeability, k, can vary by many orders of magnitude depending on the soil type. Clays may have permeability as small as about , gravels may have permeability up to about . Layering and heterogeneity and disturbance during the sampling and testing process make the accurate measurement of soil permeability a very difficult problem.

Flow nets

Darcy's Law applies in one, two or three dimensions. In two or three dimensions, steady state seepage is described by Laplace's equation. Computer programs are available to solve this equation. But traditionally two-dimensional seepage problems were solved using and a graphical procedure known called flow nets.

One set of lines in the flow net are in the direction of the water flow (flow lines), and the other set of lines are in the direction of constant total head (equipotential lines). Flow nets may be used for example to estimate the quantity of seepage under dams and sheet piling. A plan flow net to estimate flow of water from a stream to a discharging well

Seepage forces and erosion

When the seepage velocity is great enough, erosion can occur because of the frictional drag exerted on the soil particles. Vertically upwards seepage is a source of danger on the downstream side of sheet piling and beneath the toe of a dam or levee. Erosion of the soil, known as "piping", can lead to failure of the structure and to sinkhole formation. Seeping water removes soil, starting from the exit point of the seepage, and erosion advances upgradient. The term sand boil is used to describe the appearance of the discharging end of an active soil pipe.

Seepage pressures

Seepage in an upward direction reduces the effective stress within the soil. When the water pressure at a point in the soil is equal to the total vertical stress at that point, the effective stress is zero and the soil has no frictional resistance to deformation. For a surface layer, the vertical effective stress becomes zero within the layer when the upward hydraulic gradient is equal to the critical gradient. At zero effective stress soil has very little strength and layers of relatively impermeable soil may heave up due to the underlying water pressures.

The loss in strength due to upward seepage is a common contributor to levee failures. The condition of zero effective stress associated with upward seepage is also called liquefaction, Quicksand, or a boiling condition. Quicksand was so named because the soil particles move around and appear to be 'alive' (the biblical meaning of 'quick' - as opposed to 'dead').

Reservoirs store surface water for use at a later time. Surface water reservoirs are created by building a dam or river at the outlet end of a LAKE or reservoir to store WATER. Reservoir water may be used for both off-stream (consumptive) uses such as irrigation, municipal and industrial; and for in-stream (non-consumptive) uses such as power, recreational developments, FLOOD control and wildlife habitat improvement. Reservoirs may be used to improve the holding capacity of an existing lake or to create new lakes on rivers, creeks or coulees.

Storage Capacity

Reservoirs, as discussed here, do not include any type of subsurface reservoir structure that stores water, natural gas or oil. Surface water reservoirs are usually classified by 3 basic physical characteristics: maximum depth, flooded area and the volume of water stored at their full supply level (FSL). Large reservoirs may have several target storage levels, each level planned to ensure a usable amount of water at a certain time of year. For instance, in addition to its normal or FS level, a reservoir may have a conservation storage level (ie, the desirable level for a flood-control reservoir at the start of each flooding period), a flood storage level (the maximum desirable level that is permitted for reservoir safety during floods), a dead storage level (the level below which water may not be withdrawn for consumptive uses) and various other target levels established to meet such needs as summer recreational use, irrigation uses and hydropower discharges throughout the year.

Before construction, reservoir sites are subjected to detailed studies to determine their ability to meet the needs of potential users and to assess the overall benefits that can be derived from the project. Studies are usually carried out on a monthly basis, using extensive periods of estimated or recorded monthly historical streamflow to evaluate the project's performance under DROUGHT and flood conditions that have occurred historically. The results are used to determine if proposed reservoir would adequately serve its potential users if similar droughts occurred in the future.

A reservoir used for municipal purposes will be designed to have no shortages because the users (people in cities) cannot tolerate periods without water. When uses are such that shortages, although they may be inconvenient, will not cause severe economic hardship, reservoirs are often designed to allow for an acceptable percentage of shortages in critically dry years. This permits more effective use to be made of the available water resources. Two or more reservoirs are often used to maximize the supply potential of a DRAINAGE BASIN. The potential improvement that might be achieved by adding additional potential reservoir storage sites to an existing system may be evaluated by hydrologic studies involving both the existing and the proposed reservoirs. The studies assume that they are being utilized in combination to optimize their overall effectiveness.

Evaporation

Water evaporates from the surface of reservoirs when they are not covered by ICE. These losses, known as net evaporation, may be estimated by subtracting measured or estimated precipitation from estimated evaporation on a reservoir's surface. The volume of reservoir losses caused by net evaporation may be calculated by multiplying net evaporation by a reservoir's surface area.

Water Cycle

The water cycle is a key element of change in the environment and is essential to all life (artwork by Michael Lee).

When hydrologic studies indicate that the addition of further storage capacity will not significantly increase the yield of a reservoir, either because the additional water stored would evaporate before being used or because all available water has been used, the reservoir has reached its optimum storage capacity and further storage increases would be wasted. It may by feasible to add storage for flood-control purposes. This alternative would require additional studies of potential flooding problems in the downstream river basin to determine if such storage is warranted.

Water Discharge

Reservoirs have spillways to divert excess water during above-normal periods. The spillway's capacity is based both on the economic consequences and on the potential danger to human life that might occur if the dam fails. When the potential exists for downstream loss of life, extensive property damage or both, spillways are usually designed to carry the largest flood that may be expected to occur at that site. After construction, reservoirs are operated to co-ordinate consumptive, recreational and power uses with downstream releases to maintain predetermined levels of minimum flow, thus establishing monthly reservoir levels, regulated discharge and diversions.

Environmental Effects

Several changes occur to the environment adjacent to a reservoir. The lake created by the reservoir provides an additional recreational area, but the pre-existing river is destroyed and the area at the river's edge that used to provide shelter to animals and birds is lost. The river downstream from the reservoir usually has lower peaks and higher low flows because of the moderating effect of the reservoirs during floods and because of additional water released during dry periods.

Colder water is released from the reservoir and fish must either adjust to the new environment or emigrate. However, there is often an increase in volume and diversity of the fish stocked in the reservoir itself, resulting in a net gain to the fishery.

The overall water quality in the reservoir and downstream tends to improve because the reservoir acts as a mixing bowl, averaging out normal seasonal fluctuations. The clarity of downstream water also improves as the river's sediment load is trapped by the reservoir. This process may have an adverse effect on the channel below the reservoir both because the streambed erodes and because the natural environment that contributed to the downstream wildlife and vegetation is disturbed. It no longer derives benefits from the annual flooding cycle of sediment-laden water. Similarly, at the upstream end of the reservoir, the capacity of the inlet channel is decreased by sedimentation.

LAKE DIEFENBAKER is a good example of a multi-use reservoir. After 100 years of proposals and plans the project was approved in 1958. On 25 May 1959 an inaugural ceremony marked the start of construction on the project's main dam, GARDINER DAM on the South Saskatchewan River. This dam, and the Coteau Creek generating station, designed to develop hydroelectric power at the damsite, were completed in 1967. A second dam was built on the QU'APPELLE RIVER to keep water from escaping down that stream. In the spring of 1968 Lake Diefenbaker filled to operating capacity and the first power was generated. In 1970, the reservoir filled to its full capacity for the first time.

At its full supply level the dam controls a reservoir that stores a total of 9.4 billion m3, but only 4 billion m3 of this is live capacity. The remaining 5.4 billion m3 is dead storage, and cannot be used for power or to release water downstream. The maximum depth of water at the dam is 58 m and the reservoir created by Gardiner Dam is 225 km long with a shoreline of 760 km. The spillway is 161 m wide and can pass, at its peak capacity, 7500 m3 of water per second. In an average year 250 000 m3 of water is lost to evaporation from the lake's surface area of 430 km2.

Lake Diefenbaker water is also used for irrigation purposes. In 1986 some 16 000 ha were irrigated, and by 1997 that figure had increased to 43 000 ha. It is estimated that, by the year 2030, 115 000 ha will be irrigated. The Coteau Creek generating station develops an average of 775 million kWh each year, and the downstream hydropower stations in the basin benefit directly from the flow regulation of Lake Diefenbaker. For instance, the Squaw Rapids Plant on the Saskatchewan River near Nipawin, Sask, is now able to generate 40% of its power in the winter (double the previous amount) owing directly to improved outflow from Lake Diefenbaker. Water from the lake is also used to supply 10 potash mines, 11 communities and a series of recreational and wildlife projects using water diverted from the lake.

The lake itself has become a water-based resort area, creating a major recreational lake in a traditionally dry area of Saskatchewan, and waters from this lake are annually diverted into the Qu'Appelle River system to augment the municipal water supplies of Moose Jaw and Regina and to improve the quality and quantity of water in the Qu'Appelle River basin.

6

DRIP IRRIGATION SYSTEM

Drip irrigation is the most efficient water and nutrient delivery system for growing crops. It delivers water and nutrients directly to the plant's roots zone, in the right amounts, at the right time, so each plant gets exactly what it needs, when it needs it, to grow optimally. Thanks to drip irrigation, farmers can produce higher yields while saving on water as well as fertilizers, energy and even crop protection products. Drip irrigation or trickle irrigation is a type of micro-irrigation system that has the potential to save water and nutrients by allowing water to drip slowly to the roots of plants, either from above the soil surface or buried below the surface. The goal is to place water directly into the root zone and minimize evaporation. Drip irrigation systems distribute water through a network of valves, pipes, tubing, and emitters. Depending on how well designed, installed, maintained and operated it is, a drip irrigation system can be more efficient than other types of irrigation systems, such as surface irrigation or sprinkler irrigation.

Drip Irrigation method, also called drop by drop irrigation, and underground irrigation, functions as its name imply. Drop by drop, water is distributed to the active root zone of plants. This method, if managed properly, might be the most water-supply-efficient way of irrigation, because runoff and evaporation reduced significantly. Drip irrigation in today's agriculture, is frequently integrated with plastic sheet, further diminishing evaporation, and is also a method of fertilizers delivery to the plants. This process called fertigation (fertilizer + irrigation). DIY DripKits, complete professional dripsystems.

Deep trickle (percolation), where water moves below the active root zone, can happen if a drip system is running for too long time or if the irrigation application rate is too high. Drip irrigation systems vary from high-end and computer controlled to simple low-tech with intensive labor involved. Relatively low pressures, is required for most types of micro irrigation systems. System can be planned with uniformity within the field or for exact water conduction to individual plants in a field containing a mix of plant types.

Even though it is complex to regulate water pressure on steep slopes, PC emitters (pressure compensating) available, so drip systems can handle you can handle un-leveled fields as well. Water pressure regulation and filtration to remove particles are primordial in any drip system. Drip systems tubes are usually black, however there are also brown, blue and even white tubes. Some drip tubes are buried under soil or mulch to prevent the growth of algae and to protect the PE (polyethylene) from degradation due to ultra-violet radiation. Sometimes a drip irrigation system can also be a porous clay vessel inserted into the soil and periodically filled from a hose or a watering can. Successful farmers around the globe use drip systems to irrigate almost any intensive crop or plant. find more information on drip irrigation equipment.

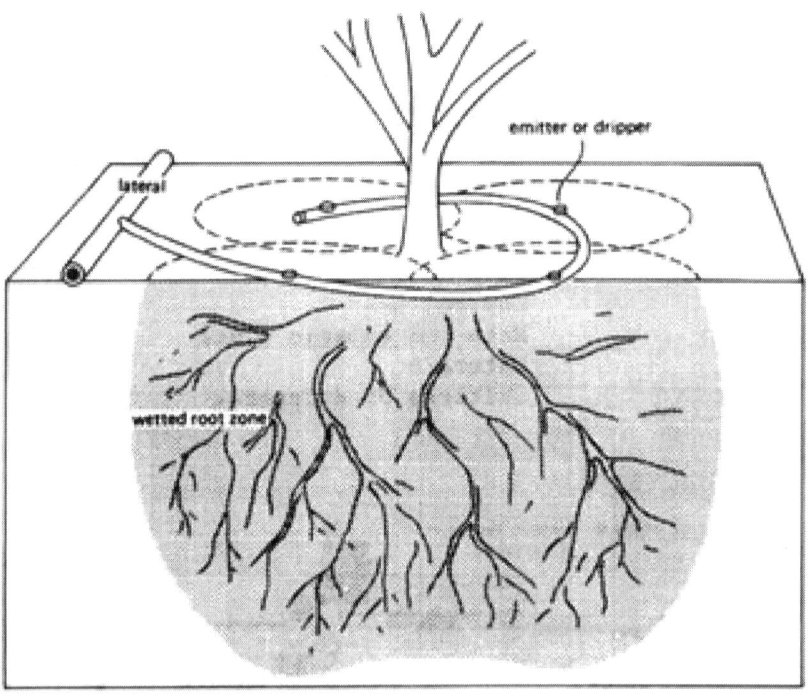

Figure : drip irrigation, only the part of the soil in which the roots grow is wetted

Latest drip irrigation technology called hydro-pc, guarantee a wide, efficient labyrinth leading the water into a flow control cell, where a special diaphragm compensate the stream, and keeps a stable flow rate at changeable inlet pressures. IG hydropc drip line is highly resistant to clogging. The dripper inlet filter, located closer to the tube center, guarantees an excellent constant flushing treatment. The relatively small diameter drip line makes a better drip irrigation efficiency. Decrease retained water in the system, causing a shorter refilling time with less redundant drained water, on sloppy terrains, no excessive water wetting in the lower parts of the field. Competitive low cost and big saving on transportation volume and weight. Applications for hydro-pc: All kind of crops on sloping terrain. Or where long run drip lines are a must, look at our irrigation systems design section.

Drip irrigation is sometimes called trickle irrigation and involves dripping water onto the soil at very low rates (2-20 litres/hour) from a system of small diameter plastic pipes fitted with outlets called emitters or drippers. Water is applied close to plants so that only part of the soil in which the roots grow is wetted (Figure 60), unlike surface and sprinkler irrigation, which involves wetting the whole soil profile.

With drip irrigation water, applications are more frequent (usually every 1-3 days) than with other methods and this provides a very favourable high moisture level in the soil in which plants can flourish.

Expensive commercial drip irrigation systems are employed in highly technical and industrial farming. The used systems are very expensive and needs expert design and maintenance.

Most large drip irrigation systems employ some type of filter to prevent clogging of the small emitter flow path by small waterborne particles. New technologies are now being offered that minimise clogging. Some residential systems are installed without additional filters since potable water is already filtered at the water treatment plant. Virtually all drip irrigation equipment manufacturers recommend that filters be employed and generally will not honour warranties unless this is done. Last line filters just before the final delivery pipe are strongly recommended in addition to any other filtration system due to fine particle settlement and accidental insertion of particles in the intermediate lines. Drip systems often mix liquid fertiliser with the irrigation water. This is called fertigation and chemigation (application of pesticides and other chemicals to periodically clean out the system).

Small Scale and Self-Made Drip Irrigation Systems

For a relatively low initial investment a small-scale farmer can buy and set up a drip-irrigation system. If used to grow crops for market, this investment will pay itself within the first season and lead to increased household food production, especially during extended dry periods. Apply correct design (that might need training of the farmers), very simple drip systems can be built with local available material. Using buckets or barrels as water reservoir and bamboo or PVC tubes as distribution pipes, everyone can construct a very efficient irrigation system. If wastewater is used, a filtration unit after the treatment plant is recommended to avoid clogging of the emitters.

Design of a Simple Drip Irrigation System

A simple drip can consist of a 20 litre bucket with 30 metres (100 feet) of hose or drip tape connected to the bottom of the tank. The bucket is placed at least 1 metre (3 feet) above the ground so that gravity provides sufficient water pressure to ensure even watering for the entire crop. Clean water is poured into the bucket daily through a filter/ strainer. The water in the bucket fills the drip tape and is evenly distributed to 100 watering points. A multi-chambered plastic drip tape is engineered to dispense water through openings spaced at 30cm (12 inches). The bucket kit is the smallest type of drip irrigation technique (Adapted from RCSD 2008). A filter after the control valve can be installed, to prevent blockages (e.g. a screen) or an even more developed filter to improve the water quality.

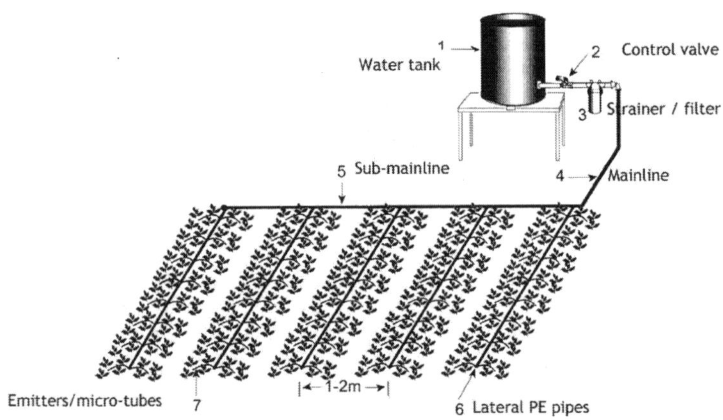

Costs considerations

As already mentioned, commercial systems for industrial production are very expensive. Small scale farmers can buy a drip irrigation system for relatively low initial costs (US$15 to $85) or construct it with local available material (buckets, bamboo or plastic pips). In general, it is more costly than manual irrigation, but has improved yields and decreased water/ operating costs.

OPERATION AND MAINTENANCE

For perennial crops, the drip hose should be lifted periodically if a drip hose system is used on the soil surface, so that leaves, soil, and debris do not cover the hose. If the drip hose is not lifted, roots can grow over the hose, anchor it to the ground, and eventually pinch off the flow of water. Leaks can occur unexpectedly as a result of damage by insects, animals, or farming tools. Systematically monitor the lines for physical damage. It is important to fix holes as soon as possible to prevent uneven irrigation. If the rate of water flow progressively declines during the season, the tubes or tape may be slowly plugging, resulting in severe damage to the crop. Once a month, flush the drip lines by opening the far ends of a portion of the tubes at a time and allowing the higher velocity water to flush out the sediment. If poorly treated wastewater is used, soil quality can be degraded over time (e.g. accumulation of salts) (TILLEY at al. 2014).

Health Aspects

If wastewater is used for the irrigation process, there are potential health risks if water is not properly pre-treated (i.e. inadequate pathogen reduction). If poorly treated wastewater is applied. Appropriate pre-treatment should precede any irrigation scheme to limit health risks to those who come in contact with the water. As well, depending on the degree of treatment that the effluent has undergone, it may be contaminated with the different chemicals that are discharged into the system. When effluent is used for irrigation, households and industries connected to the system should be made aware of the products that are and are not appropriate for discharging into the system. Drip irrigation is the only type of irrigation that should be used with edible crops, and even then, care should be taken to prevent workers and harvested crops from coming in contact with the treated effluent. Despite safety concerns, irrigation with effluent is an effective way to recycle nutrients and water (see also waterborne diseases pathogens and contaminants).

Components and operation

Components used in drip irrigation (listed in order from water source) include:

Pump or pressurized water source

- Water filter(s) or filtration systems: sand separator, Fertigation systems (Venturi injector) and chemigation equipment (optional)
- Backwash controller (Backflow prevention device)
- Pressure control valve (pressure regulator)
- Distribution lines (main larger diameter pipe, maybe secondary smaller, pipe fittings)
- Hand-operated, electronic, or hydraulic control valves and safety valves
- Smaller diameter polyethylene tube (often called "laterals")

- Poly fittings and accessories (to make connections)
- Emitting devices at plants (emitter or dripper, micro spray head, inline dripper or inline drip tube)
- In drip irrigation systems, pump and valves may be manually or automatically operated by a controller.

Most large drip irrigation systems employ some type of filter to prevent clogging of the small emitter flow path by small waterborne particles. New technologies are now being offered that minimize clogging. Some residential systems are installed without additional filters since potable water is already filtered at the water treatment plant. Virtually all drip irrigation equipment manufacturers recommend that filters be employed and generally will not honor warranties unless this is done. Last line filters just before the final delivery pipe are strongly recommended in addition to any other filtration system due to fine particle settlement and accidental insertion of particles in the intermediate lines.

Drip and subsurface drip irrigation is used almost exclusively when using recycled municipal wastewater. Regulations typically do not permit spraying water through the air that has not been fully treated to potable water standards.

Because of the way the water is applied in a drip system, traditional surface applications of timed-release fertilizer are sometimes ineffective, so drip systems often mix liquid fertilizer with the irrigation water. This is called fertigation; fertigation and chemigation (application of pesticides and other chemicals to periodically clean out the system, such as chlorine or sulfuric acid) use chemical injectors such as diaphragm pumps, piston pumps, or aspirators. The chemicals may be added constantly whenever the system is irrigating or at intervals. Fertilizer savings of up to 95% are being reported from recent university field tests using drip fertigation and slow water delivery as compared to timed-release and irrigation by micro spray heads.

Properly designed, installed, and managed, drip irrigation may help achieve water conservation by reducing evaporation and deep drainage when compared to other types of irrigation such as flood or overhead sprinklers since water can be more precisely applied to the plant roots. In addition, drip can eliminate many diseases that are spread through water contact with the foliage. Finally, in regions where water supplies are severely limited, there may be no actual water savings, but rather simply an increase in production while using the same amount of water as before. In very arid regions or on sandy soils, the preferred method is to apply the irrigation water as slowly as possible.

Pulsed irrigation is sometimes used to decrease the amount of water delivered to the plant at any one time, thus reducing runoff or deep percolation. Pulsed systems are typically expensive and require extensive maintenance. Therefore, the latest efforts by emitter manufacturers are focused on developing new technologies that deliver irrigation water at ultra-low flow rates, i.e. less than 1.0 L (2.1 US pints; 1.8 imperial pints) per hour. Slow-and-even delivery further improves water use efficiency without incurring the expense and complexity of pulsed delivery equipment.

An emitting pipe is a type of drip irrigation tubing with emitters pre-installed at the factory with specific distance and flow per hour as per crop distance.

An emitter restricts water flow passage through it, thus creating head loss required (to the extent of atmospheric pressure) to emit water in the form of droplets. This head loss is achieved by friction/turbulence within the emitter.

SUITABLE CROPS

Drip irrigation is most suitable for row crops (vegetables, soft fruit), tree and vine crops where one or more emitters can be provided for each plant. Generally only high value crops are considered because of the high capital costs of installing a drip system.

Suitable Slopes

Drip irrigation is adaptable to any farmable slope. Normally the crop would be planted along contour lines and the water supply pipes (laterals) would be laid along the contour also. This is done to minimize changes in emitter discharge as a result of land elevation changes.

Suitable Soils

Drip irrigation is suitable for most soils. On clay soils water must be applied slowly to avoid surface water ponding and runoff. On sandy soils higher emitter discharge rates will be needed to ensure adequate lateral wetting of the soil.

Suitable Irrigation Water

One of the main problems with drip irrigation is blockage of the emitters. All emitters have very small waterways ranging from 0.2-2.0 mm in diameter and these can become blocked if the water is not clean. Thus it is essential for irrigation water to be free of sediments. If this is not so then filtration of the irrigation water will be needed.

Blockage may also occur if the water contains algae, fertilizer deposits and dissolved chemicals which precipitate such as calcium and iron. Filtration may remove some of the materials but the problem may be complex to solve and requires an experienced engineer or consultation with the equipment dealer.

Drip irrigation is particularly suitable for water of poor quality (saline water). Dripping water to individual plants also means that the method can be very efficient in water use. For this reason it is most suitable when water is scarce.

Drip System Layout

A typical drip irrigation system is shown and consists of the following components:

Pump unit

Control head

Main and submain lines

Laterals

Emitters or drippers.

The pump unit takes water from the source and provides the right pressure for delivery into the pipe system.

The control head consists of valves to control the discharge and pressure In the entire system. It may also have filters to clear the water. Common types of filter include screen filters and graded sand filters which remove fine material suspended in the water. Some control head units contain a fertilizer or nutrient tank. These slowly add a measured dose of fertilizer into the water during irrigation. This is one of the major advantages of drip irrigation over other methods.

Mainlines, submains and laterals supply water from the control head into the fields. They are usually made from PVC or polyethylene hose and should be buried below ground because they easily degrade when exposed to direct solar radiation. Lateral pipes are usually 13-32 mm diameter.

Emitters or drippers are devices used to control the discharge of water from the lateral to the plants. They are usually spaced more than 1 metre apart with one or more emitters used for a single plant such as a tree. For row crops more closely spaced emitters may be used to wet a strip of soil. Many different emitter designs have been produced in recent years. The basis of design is to produce an emitter which will provide a specified constant discharge which does not vary much with pressure changes, and does not block easily.

Operating Drip Systems

A drip system is usually permanent. When remaining In place during more than one season, a system is considered permanent. Thus it can easily be automated. This is very useful when labour is scarce or expensive to hire. However, automation requires specialist skills and so this approach is unsuitable if such skills are not available.

Water can be applied frequently (every day if required) with drip irrigation and this provides very favourable conditions for crop growth. However, if crops are used to being watered each day they may only develop shallow roots and If the system breaks down, the crop may begin to suffer very quickly.

Wetting patterns

Unlike surface and sprinkler irrigation, drip irrigation only wets part of the soil root zone. This may be as, low as 30% of the volume of soil wetted by the other methods. The wetting patterns which develop from dripping water onto the soil depend on discharge and soil type.

The effect of changes in discharge on two different soil types, namely sand and clay.

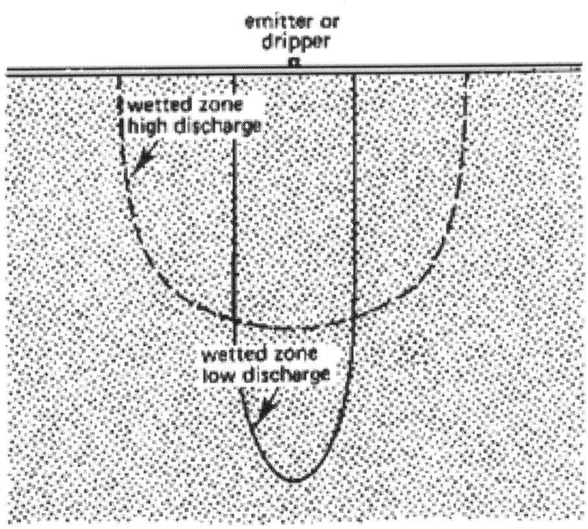

Figure: Wetting patterns for sand and clay soils with high and low discharge rates (SAND)

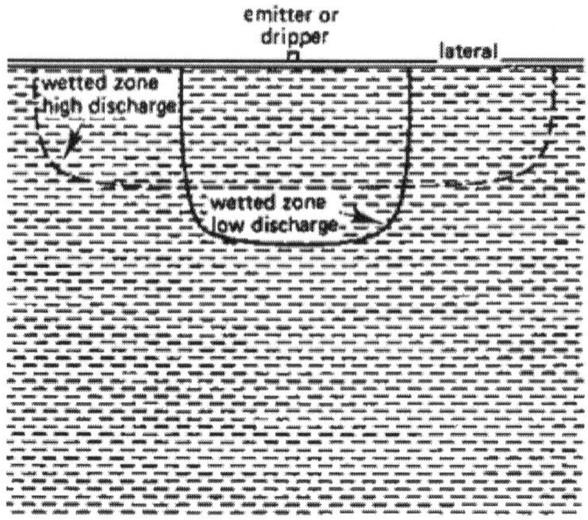

Figure: Wetting patterns for sand and clay soils with high and low discharge rates (CLAY)

Although only part of the root zone is wetted it is still important to meet the full water needs of the crop. It is sometimes thought that drip irrigation saves water by reducing the amount used by the crop. This is not true. Crop water use is not changed by the method of applying water. Crops just require the right amount for good growth.

The water savings that can be made using drip irrigation are the reductions in deep percolation, in surface runoff and in evaporation from the soil. These savings, it must be remembered, depend as much on the user of the equipment as on the equipment itself.

Drip irrigation is not a substitute for other proven methods of irrigation. It is just another way of applying water. It is best suited to areas where water quality is marginal, land is steeply sloping or undulating and of poor quality, where water or labour are expensive, or where high value crops require frequent water applications.

Bamboo drip irrigation system

Dating back 200 years, tribes in northeast India have used bamboo drip irrigation as a means of bringing water to seasonal crops. This timeless and traditional technology uses locally available material while harnessing the forces of gravity. An assortment of holed bamboo shoots zig-zag downhill, diverting the natural flow of streams and springs across terraced cropland. The advantages of using bamboo are two-fold: it prevents leakage, increasing crop yield with less water, and makes use of natural, local, and inexpensive material.

The Jaintia, Khasi, and Garo hills of Meghalaya are largely made up of steep slopes and generally rocky terrain where the soil has low water retention capacity and where the use of groundwater channels is impossible. During the dry seasons, rainfed crops such as paddy, betal leaf, and black peppers can be irrigated by bamboo drip irrigation.

Within the Jainta hills, the small village of Nongbah relies on terrace agriculture for paddy cultivation. There are no restrictions for individuals tapping into water flows from perennial streams, natural springs, or collection ponds. This enables farmers, nearly 97% of the population, to cultivate paddy, betal leaf, and black peppers in seasonal rotations. Meanwhile, drinking needs are met by perennial springs during the dry months, from October to March.

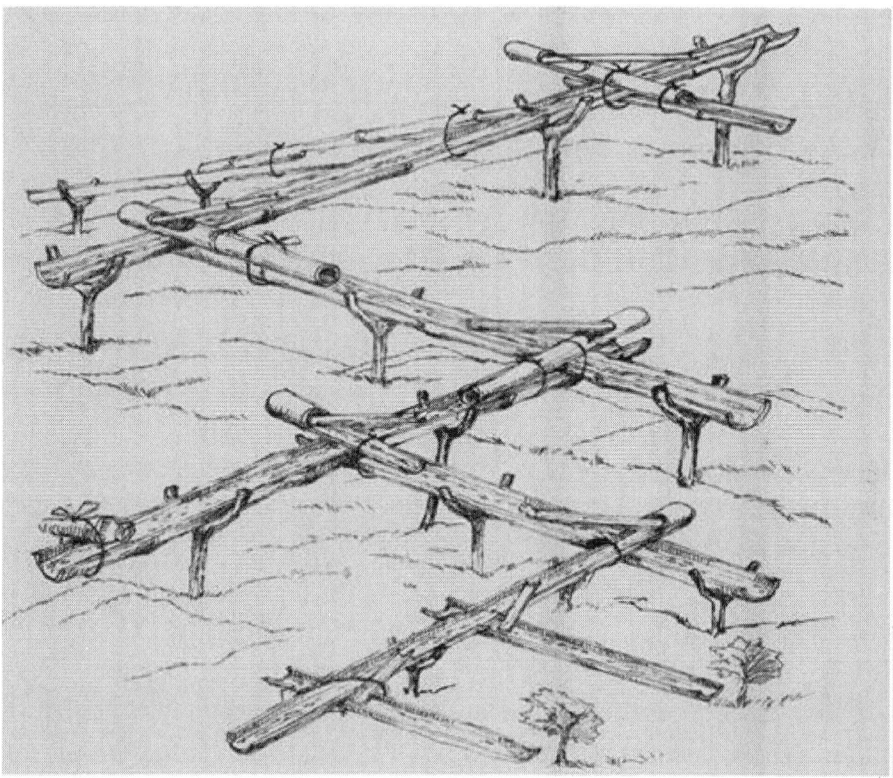

Only during the winter is irrigation required, and the bamboo system is used for crops that need relatively less water.

The few materials needed are a small dao (a type of local axe), bamboo strands of various sizes, forked branches, smaller bamboo shoots used for the channel diversions, and two willing laborers. A. Singh, in his book Bamboo Drip Irrigation Systems, investigated its use in Meghalaya and says that two workers can construct a system covering one hectare of land in 15 days. About four or five stages of irrigation zig-zag from the water source to the last point of application. Along the way, 18-20 liters of water will eventually disseminate at a rate of 20-80 drops per minute.

To get started, locate an available water source. Next, select a sloped area of land (at least 30 meters in variation). Then slice the bamboo shoots and forked branches, placing the wider shoots in the first channel and the smaller pipes for the last section (plan for 5 stages). Puncture a series of holes in the shoots, spacing them equally. Ground clearance should progressively descend so that the water may be dropped near the roots of plants in the last section (10-15 cm above ground). To reinforce the structure, tie the pipes and forked branches together using fiber-rich twine as rope. At points of diversion, smaller bamboo shoots may be used to redirect water.

Materials used during installation last around three years, while maintenance is limited to cleaning and reinforcement after seasonal monsoons. Cost is also limited to labor, which can be carried out by farmers themselves. This traditional water structure is indeed contextual to location. Such variables include: replenishable supplies of bamboo, upland sources of water, and the presence of traditional terrace agriculture.

Adapting to drier growing seasons, farmers are advised to match irrigation decisions with crop selection. It is suggested that during the June to September season, rice cultivation can be stopped and farmers can opt for crops such as pulses, sunflower, sesame and maize. In areas affected severely by drought, farmers may go for pearl millet, minor millet and forage crops. However, in places where there is water stagnation, they can continue to grow rice.

In the Meghalaya hills there are an estimated 3,108 square kilometers of bamboo forests. In 1990, it was estimated that the total yield of bamboo in the state was 2.09 tons/hectare/year. The 38 different species of bamboo in the region are typically harvested at the community level and loosely managed by the Autonomous District Councils (ADC). Bamboo supplies have recently come under threat from a surge in rodent populations, gregarious flowering, disease, and large-scale extraction. Nonetheless, large-scale conservation and protection plans are underway in Arunachal Pradesh, Assam, Manipur, Meghalaya, Nagaland, Mizoram, Tripura, and Sikkim, which together contain over 50% of India's bamboo supply (within 226,000 square kilometers of land.)

The Jaintia, Khasi, and Garo hill tribes have long entrusted the use of bamboo drip irrigation as a means to fulfilling domestic, agricultural, and customary needs. Its function remains unspoiled so as the rains continue to fall and the bamboo continues to grow.

Drip tape

Drip tape is a type of thin-walled dripperline used in drip irrigation. The first drip tape was known as "Dew Hose".

Drip tape duct tape is made of polyethylene and is sold flat on reels. The wall thickness typically ranges from 0.1 to 0.6 mm (4 to 25 mils). Thicker walled tapes are commonly used for permanent subsurface drip irrigation and thinner walled tapes for temporary throw-away type systems in high-value crops.

Water exits from tape through emitters or drippers. The typical emitter spacing ranges from 150 to 610 mm (6 to 24 inches). In some products, the emitters are manufactured simultaneously with the tape and are actually formed as part of the product itself. In others, the emitters are manufactured separately and installed at the time of production.

Some product is not a tape, but a thin-walled dripperline, but in popular parlance, both types of products are called tapes. Typical tape diameters are 5/8 in (16 mm), 7/8 in (22 mm), and 1+3/8 in (35 mm), with the larger diameters more commonly used on permanent installations with longer runs.

Drip tape is a recyclable material and can be recycled into viable plastic resins for reuse in the plastics manufacturing industry.

Uses

Drip irrigation is used in farms, commercial greenhouses, and residential gardens. Drip irrigation is adopted extensively in areas of acute water scarcity and especially for crops and trees such as coconuts, containerized landscape trees, grapes, bananas, ber, eggplant, citrus, strawberries, sugarcane, cotton, maize, and tomatoes.

Drip irrigation kits for home gardens are increasingly popular for homeowners and consist of a timer, hose, and emitter. Hoses that are 4 mm (0.16 in) in diameter are used to irrigate flower pots.

- Advantages and disadvantages
- The advantages of drip irrigation are:
- Fertilizer and nutrient loss is minimized due to a localized application and reduced leaching.
- Water application efficiency is high if managed correctly.
- Field leveling is not necessary.
- Fields with irregular shapes are easily accommodated.
- Recycled non-potable water can be safely used.
- Moisture within the root zone can be maintained at field capacity.
- Soil type plays a less important role in the frequency of irrigation.
- Soil erosion is lessened.

- Weed growth is lessened.
- Water distribution is highly uniform, controlled by the output of each nozzle.
- Labour cost is less than other irrigation methods.
- Variation in supply can be regulated by regulating the valves and drippers.
- Fertigation can easily be included with minimal waste of fertilizers.
- Foliage remains dry, reducing the risk of disease.
- Usually operated at lower pressure than other types of pressurized irrigation, reducing energy costs.
- The disadvantages of drip irrigation are:
- Initial cost can be more than overhead systems.
- The sun can affect the tubes used for drip irrigation, shortening their lifespan. (See Polymer degradation);

The risks of degrading plastic affecting the soil content and food crops. With many types of plastic, when the sun degrades the plastic, causing it to become brittle, the estrogenic chemicals (that is, chemicals replicating female hormones) which would cause the plastic to retain flexibility have been released into the surrounding environment.

If the water is not properly filtered and the equipment not properly maintained, it can result in clogging or bioclogging.

For subsurface drip the irrigator cannot see the water that is applied. This may lead to the farmer either applying too much water (low efficiency) or an insufficient amount of water, this is particularly common for those with less experience with drip irrigation.

Drip irrigation might be unsatisfactory if herbicides or top dressed fertilizers need sprinkler irrigation for activation.

Drip tape causes extra cleanup costs after harvest. Users need to plan for drip tape winding, disposal, recycling or reuse.

Waste of water, time and harvest, if not installed properly. These systems require careful study of all the relevant factors like land topography, soil, water, crop and agro-climatic conditions, and suitability of drip irrigation system and its components.

In lighter soils subsurface drip may be unable to wet the soil surface for germination. Requires careful consideration of the installation depth.

Most drip systems are designed for high efficiency, meaning little or no leaching fraction. Without sufficient leaching, salts applied with the irrigation water may build up in the root zone, usually at the edge of the wetting pattern. On the other hand, drip irrigation avoids the high capillary potential of traditional surface-applied irrigation, which can draw salt deposits up from deposits below.

The PVC pipes often suffer from rodent damage, requiring replacement of the entire tube and increasing expenses.

Drip irrigation systems cannot be used for damage control by night frosts (like in the case of sprinkler irrigation systems).

7

IRRIGATION EFFICIENCY: IRRIGATION ENGINEERING

The amount of supplied water is not fully utilised for the growth of crops because there are several reasons for water loss. Irrigation efficiency is defined as the ratio of the amount of water available to the amount of water supplied from the reservoir. It is expressed in percentage.

The amount of water used for irrigation varies depending on the climate and on the crop being grown, and it also depends on the irrigation technique used. Just like in your garden or home landscaping there are more or less efficient sprinklers. In many parts of the world flood or surface, irrigation is still used where water flows across a field and soaks into the soil.

The term irrigation efficiency expresses the performance of a complete irrigation system or components of the system. Irrigation efficiency is defined as the ratio between the amount of water used to meet the consumptive use requirement of crop plus that necessary to maintain a favourable salt balance in the crop root zone to the total volume of water diverted, stored or pumped for irrigation. Thus, water applied by the irrigation system and not being made available to be taken up by plant roots is wasted and reduces irrigation efficiency.

The major causes for reduced irrigation efficiency include storage losses, conveyance losses and field application losses. In India, overall irrigation efficiency of major irrigation projects ranges between 35-40%. This is one of the reasons for increasing gap between irrigation potential created (102.77 M ha till end of 10th plan 2007) and utilized (87.23 M ha). This gap of about 16%, is same as the irrigation potential created between 1951 and 1970. At the end of eighth plan, Planning commission estimated that with a 10% increase in the present level of water use efficiency in irrigation systems, an additional 14 Mha area can be brought under irrigation from the existing irrigation capacities. In order to meet the growing demands of water for food, environment, urban and industry, it is necessary to improve irrigation efficiency at all levels.

Surface or flood irrigation is the least efficient manner of irrigation. When a field is flooded, more water than is needed by the plant is applied to the field and water evaporates, seeps into the ground and percolates down to the groundwater, where it can be out of reach of the plant's roots. Another problem with flood irrigation is that the water is not always applied evenly to all plants. Some plants might get too much water, and others get too little. On the other hand, flood irrigation tends to use the least energy of any irrigation system.

Furrow irrigation is another type of surface irrigation in which water is directed through gated pipe or siphon tubes into furrows between rows of plants. When using furrow irrigation, water is lost to surface runoff, groundwater, and evaporation, and it can be challenging to get water evenly to an entire field.

More efficient methods of irrigation include drip irrigation sprinklers, and micro-spray irrigation. All of these methods, while more efficient, also require significant investments in equipment, pipes, infrastructure and energy. In addition to the high cost, some soil types, irrigation networks, field sizes and crops pose greater challenges to implementation of more efficient methods of irrigation. For example, in the Grand Valley of western Colorado, the irrigation network is entirely gravity-fed, meaning that farmers can easily flood and furrow irrigate without the use of pumps. In addition, the fields are small and the soils are very clayey, all of which make using center pivots for row crops particularly challenging and expensive. But, in the same valley, the peach orchards have successfully used micro-spray and drip systems. A major advantage of more efficient irrigation in addition to reduced water consumption is that crop yields are often higher because the water can be applied more directly to the plant when water is needed.

WATER APPLICATION EFFICIENCY

The ability to manage an irrigation system is contingent on an accurate estimate of the percentage of water pumped that becomes available for crop use. No irrigation system delivers water at 100 percent efficiency. Water may be lost through delivery systems or pipelines and some water may remain in the soil, but not be used by the crop. Some water may run off the soil surface into lowland areas. Still other water may be lost to evaporation in the air, or from the soil and plant surfaces. Figures 8.1a and 8.1b show the major losses for sprinkler and surface irrigation systems. To know how much water to pump, these losses must be totaled and added to the amount of water needed by the crop.

In most cases, the goal is to insure that all areas of the field receive a set amount of uniformly applied water. The cans recording application depths below the horizontal line are not receiving enough water — catches are less than the desired 0.85 inches. Another application will be needed to insure that the entire field receives at least 0.85 inches of water. This will require using more water and energy than is necessary. If this pattern occurs during each irrigation, plants in the areas receiving less than 0.85 inches eventually could experience water stress. The cans recording application depths above the line receive at least 0.85 inches of water. Any extra water applied could lead to surface runoff

Water application efficiency accounts for how uniformly the water is applied and can be used for other assessments. If the center pivot owner is trying to decide whether switching to a new sprinkler package would be economical, the change in water application efficiency could be a major factor. If water becomes limited, changing to a system with a higher water application efficiency will provide more useable water to the crop and reduce pumping costs.

To maximize irrigation water use, it must be uniformly applied in the right amount and at the right time. Reaching these objectives requires knowledge of water delivery characteristics, field soils and slopes, and the expected crop water use rates.

Mathematical relationships have been developed to help quantify the amount of applied water that becomes available for plant use.

Water application efficiency refers to the amount of water applied that is stored in the crop root zone. This value is determined by water distribution characteristics, system management, soil conditions, the crop, and weather conditions. Water application efficiency pertains to an individual irrigation event.

Definition

The ratio of the amount of water available (output) to the amount of water supplied (input) is known as Irrigation Efficiency. It is expressed in percentage.

Types of Irrigation Efficiency

The following are the various types of irrigation efficiencies:

(a) Water Conveyance Efficiency (?c):

It is the ratio of the amount of water applied, to the land to the amount of water supplied from the reservoir. It is obtained by the expression,

?c = Wl Wr × 100 Where, ?c = Water conveyance efficiency Wl = Amount of water applied to land Wr = Amount of water supplied from reservoir

(b) Water Application Efficiency (?a):

It is the ratio of the water stored in root zone of plants to the water applied to the land. It is obtained by the expression,

?a = Wz Wl × 100 Where, ?a = Water application efficiency Wz = Amount of water stored in root zone Wl = Amount of water applied to land

(c) Water Use Efficiency (?u):

It is the ratio of the amount of water used to the amount of water applied. It is obtained by the expression,

?u = Wu Wl × 100 Where, ?u = Water use efficiency Wu = Amount of water used Wl = Amount of water applied to land

(d) Consumptive use Efficiency (?cu):

It is the ration of the consumptive use of water to the amount of water depleted from the root zone. It is obtained by the expression, ?cu = Cu Wp × 100 Where,

Not all water taken from a source (river, well) reaches the root zone of the plants. Part of the water is lost during transport through the canals and in the fields. The remaining part is stored in the root zone and eventually used by the plants. In other words, only part of the water is used efficiently, the rest of the water is lost for the crops on the fields that were to be irrigated.

Figure shows the irrigation water losses in canals; these are due to:

1. Evaporation from the water surface

2. Deep percolation to soil layers underneath the canals
3. Seepage through the bunds of the canals
4. Overtopping the bunds
5. Bund breaks
6. Runoff in the drain
7. Rat holes in the canal bunds

Figure 25 shows the irrigation water losses in the field; these are due to:

1. Surface runoff, whereby water ends up in the drain
2. Deep percolation to soil layers below the root zone

To express which percentage of irrigation water is used efficiently and which percentage is lost, the term irrigation efficiency is used.

The scheme irrigation efficiency (e in %) is that part of the water pumped or diverted through the scheme inlet which is used effectively by the plants. The scheme irrigation efficiency can be sub-divided into:

- The conveyance efficiency (ec) which represents the efficiency of water transport in canals, and
- The field application efficiency (ea) which represents the efficiency of water application in the field.

The conveyance efficiency (ec) mainly depends on the length of the canals, the soil type or permeability of the canal banks and the condition of the canals.

In large irrigation schemes more water is lost than in small schemes, due to a longer canal system. From canals in sandy soils more water is lost than from canals in heavy clay soils. When canals are lined with bricks, plastic or concrete, only very little water is lost. If canals are badly maintained, bund breaks are not repaired properly and rats dig holes, a lot of water is lost.

with

e = scheme irrigation efficiency (%)

ec = conveyance efficiency (%)

ea = field application efficiency (%)

A scheme irrigation efficiency of 50-60% is good; 40% is reasonable, while a scheme Irrigation efficiency of 20-30% is poor.

It should be kept in mind that the values mentioned above are only indicative values.

Irrigation Efficiency Project

In 2008, the Blackfoot Challenge initiated the Irrigation Efficiency Project with the goals of promoting efficient and sustainable water use while maintaining agricultural production in the Blackfoot Watershed. The Irrigation Efficiency Program consists of three primary components - irrigation system energy efficiency evaluations, irrigation scheduling or irrigation water management, and irrigation equipment maintenance.

Energy Efficiency Evaluations

Efficiency of an irrigation system can be thought of in two ways - efficiency of converting energy to work and water application efficiency. For this program, the Blackfoot Challenge uses the energy efficiency evaluation method as it covers the entire system and allows for assessment at many levels. Some of the issues commonly identified during these evaluations include worn pump components, motors, and sprinkler equipment as well as improperly sized pumps, motors, and sprinklers. Ensuring irrigation equipment is sized properly and working as efficiently as possible makes for the most efficient use of water and energy, improves crop uniformity and production, and most importantly, saves money.

Irrigators in the Blackfoot Watershed are eligible for rebates and incentives offered through Missoula Electric Cooperative and Northwestern Energy for energy saving measures. Examples of energy saving measures include rebuilding or replacing pumps, installing variable frequency drives, and sprinkler packages.

Irrigation Scheduling and Water Management

Irrigation scheduling is simply applying the right amount of water at the right time to maximize production and minimize adverse environmental impacts. When applied, irrigation scheduling has the potential to conserve energy and water, reduce production costs, and improve crop production. In addition, irrigation scheduling gives irrigators tools for adapting to changes in water availability. The Blackfoot Irrigation Scheduling Program is designed to provide area irrigators with information to better use their water and energy resources. The Program includes intensive irrigation scheduling for individual irrigators as well as web and email based information for the community at large.

Irrigation Equipment Maintenance

The efficiency evaluations will identify issues with irrigation equipment.

Unfortunately, by the time issues are identified the equipment needs to be replaced. Proper maintenance of irrigation systems can save energy and water, reduce breakdowns and repairs, lower operating costs, improve production and lengthen the life of equipment. Proper maintenance is the simplest thing an irrigator can do to conserve energy and water resources, and save money.

8

IRRIGATION ENGINEERING: WATER RESOURCE

The amount of water used for irrigation varies depending on the climate and on the crop being grown, and it also depends on the irrigation technique used. Just like in your garden or home landscaping there are more or less efficient sprinklers. In many parts of the world flood or surface, irrigation is still used where water flows across a field and soaks into the soil.

Surface or flood irrigation is the least efficient manner of irrigation. When a field is flooded, more water than is needed by the plant is applied to the field and water evaporates, seeps into the ground and percolates down to the groundwater, where it can be out of reach of the plant's roots. Another problem with flood irrigation is that the water is not always applied evenly to all plants. Some plants might get too much water, and others get too little. On the other hand, flood irrigation tends to use the least energy of any irrigation system.

Furrow irrigation is another type of surface irrigation in which water is directed through gated pipe or siphon tubes into furrows between rows of plants. When using furrow irrigation, water is lost to surface runoff, groundwater, and evaporation, and it can be challenging to get water evenly to an entire field.

Irrigation water is supplied to supplement the water available from rainfall, soil moisture storage and capillary rise. However, in many cases, it is not possible to meet the full crop water requirement throughout the season due to limited water availability. In such cases, deficit irrigation is provided in the form of life saving or supplemental irrigation. Besides meeting the crop water requirement, irrigation is also provided for field preparation, climate control (crop cooling and frost control), and leaching of excessive salts.

Irrigation is very ancient practice and can be traced with the beginning of human civilization. Importance of irrigation in agriculture is very well documented in NamradaSmriti XI, 9 which states that "no grain is ever produced without water, but too much water tends to spoil the grain; an inundation is as injurious to crop growth as a dearth of water." Hence, irrigation is essential.

Water resources are natural resources of water that are potentially useful for humans, for example as a source of drinking water supply or irrigation water. 97% of the water on the Earth is salt water and only three percent is fresh water; slightly over two thirds of this is frozen in glaciers and polar ice caps. The remaining unfrozen freshwater is found mainly as groundwater, with only a small fraction present above ground or in the air. Natural sources of fresh water include surface water, under river flow, groundwater and frozen water. Artificial sources of fresh water can include treated wastewater (wastewater reuse) and desalinated seawater. Human uses of water resources include agricultural, industrial, household, recreational and environmental activities.

Water resources are under threat from water scarcity, water pollution, water conflict and climate change. Fresh water is a renewable resource, yet the world's supply of groundwater is steadily decreasing, with depletion occurring most prominently in Asia, South America and North America, although it is still unclear how much natural renewal balances this usage, and whether ecosystems are threatened. The framework for allocating water resources to water users (where such a framework exists) is known as water rights.

Water Resource Engineering is a specific kind of civil engineering that involves the design of new systems and equipment that help manage human water resources. Some of the areas Water Resource Engineers touch on are water treatment facilities, underground wells, and natural springs. Water Resources Engineering focuses on the use and management of land and water resources in rural and urban watersheds.

Irrigation water can come from groundwater (extracted from springs or by using wells), from surface water (withdrawn from rivers, lakes or reservoirs) or from non-conventional sources like treated wastewater, desalinated water, drainage water, or fog collection. A special form of irrigation using surface water is spate irrigation, also called floodwater harvesting. In case of a flood (spate), water is diverted to normally dry river beds (wadis) using a network of dams, gates and channels and spread over large areas. The moisture stored in the soil will be used thereafter to grow crops. Spate irrigation areas are in particular located in semi-arid or arid, mountainous regions. While floodwater harvesting belongs to the accepted irrigation methods, rainwater harvesting is usually not considered as a form of irrigation. Rainwater harvesting is the collection of runoff water from roofs or unused land and the concentration of this.

Around 90% of wastewater produced globally remains untreated, causing widespread water pollution, especially in low-income countries. Increasingly, agriculture uses untreated wastewater as a source of irrigation water. Cities provide lucrative markets for fresh produce, so are attractive to farmers. However, because agriculture has to compete for increasingly scarce water resources with industry and municipal users (see Water scarcity below), there is often no alternative for farmers but to use water polluted with urban waste, including sewage, directly to water their crops. Significant health hazards can result from using water loaded with pathogens in this way, especially if people eat raw vegetables that have been irrigated with the polluted water. The International Water Management Institute has worked in India, Pakistan, Vietnam, Ghana, Ethiopia, Mexico and other countries on various projects aimed at assessing and reducing risks of wastewater irrigation. They advocate a 'multiple-barrier' approach to wastewater use, where farmers are encouraged to adopt various risk-reducing behaviours.

These include ceasing irrigation a few days before harvesting to allow pathogens to die off in the sunlight, applying water carefully so it does not contaminate leaves likely to be eaten raw, cleaning vegetables with disinfectant or allowing fecal sludge used in farming to dry before being used as a human manure. The World Health Organization has developed guidelines for safe water use.

There are numerous benefits of using recycled water for irrigation, including the low cost (when compared to other sources, particularly in an urban area), consistency of supply (regardless of season, climatic conditions and associated water restrictions), and general consistency of quality. Irrigation of recycled wastewater is also considered as a means for plant fertilization and particularly nutrient supplementation. This approach carries with it a risk of soil and water pollution through excessive wastewater application. Hence, a detailed understanding of soil water conditions is essential for effective utilization of wastewater for irrigation.

In countries where humid air sweeps through at night, water can be obtained by condensation onto cold surfaces. This is practiced in the vineyards at Lanzarote using stones to condense water. Fog collectors are also made of canvas or foil sheets. Using condensate from air conditioning units as a water source is also becoming more popular in large urban areas.

Within the European Union (EU) agriculture represents around 30% of total water abstraction. The intensity of irrigation in different countries obviously varies depending on the climate, the crops cultivated and the farming methods.

For example, the role of irrigation is completely different in Southern European countries, where irrigation is essential for agricultural production, compared to Central and Western Europe.

In fact the major part of irrigated land in Europe is located in the South with Spain, Italy, France, Greece and Portugal accounting for 85% of the total irrigated area in the EU. For example, in Spain irrigated agriculture accounts for 56% of total agricultural production, occupying only 18% of the total agricultural surface.

Water used for agriculture comes from natural or other alternative sources.

Natural sources includes rainwater and surface water (lakes and rivers). These resources must be used in a sustainable way.

Rain water resources rely on the atmospheric conditions of the area. Surface water is a limited resource and normally requires the construction of dams and reservoirs with a significant environmental impact.

Alternative sources of irrigation water are the reuse of municipal wastewater and drainage water.

However the use of recycled water for irrigation may have some adverse impacts on the public health and the environment. This will depend on the recycled water application, soil characteristics, climate conditions and agronomic practises. Therefore it is important that all these factors are taken into account in the management of recycled water.

The quality of the irrigation water may affect both crop yields and soil physical conditions, even if all other conditions and cultural practices are favorable/optimal. In addition, different crops require different irrigation water qualities.

Therefore, testing the irrigation water prior to selecting the site and the crops to be grown is critical.

The quality of some water sources may change significantly with time or during certain periods (such as in dry/rainy seasons), so it is recommended to have more than one sample taken, in different time periods.

The parameters which determine the irrigation water quality are divided to three categories: chemical, physical and biological. In this review, the chemical properties of the irrigation water are discussed.

The chemical characteristics of irrigation water refer to the content of salts in the water as well as to parameters derived from the composition of salts in the water; parameters such as EC/TDS (Electrical Conductivity/ Total Dissolved Solids), SAR (Sodium Adsorption Ratio) alkalinity and hardness.

REUSE OF WATER FOR IRRIGATION

Water reuse for irrigation is a normal practice worldwide. In Europe, for example there is a large project in Clermont-Ferrand, France since 1997 where more than 10.000m/day of tertiary treated urban wastewater are reused for irrigation of 700Ha of maize. In Italy more than 4000 Ha of various crops are irrigated with recycled water. Spain also counts with several similar projects.

The water quality used for irrigation is essential for the yield and quantity of crops, maintenance of soil productivity, and protection of the environment. For example, the physical and mechanical properties of the soil, ex. soil structure (stability of aggregates) and permeability, are very sensitive to the type of exchangeable ions present in irrigation waters.

Types of irrigation

There are several methods of irrigation. They vary in how the water is supplied to the plants. The goal is to apply the water to the plants as uniformly as possible, so that each plant has the amount of water it needs, neither too much nor too little.

Surface Irrigation

Surface irrigation is the oldest form of irrigation and has been in use for thousands of years. In surface (furrow, flood, or level basin) irrigation systems, water moves across the surface of an agricultural lands, in an order to wet it and infiltrate into the soil. Surface irrigation can be subdivided into furrow, borderstrip or basin irrigation. It is often called flood irrigation when the irrigation results in flooding or near flooding of the cultivated land. Historically, this has been the most common method of irrigating agricultural land and still used in most parts of the world. Where water levels from the irrigation source permit, the levels are controlled by dikes, usually plugged by soil. This is often seen in terraced rice fields (rice paddies), where the method is used to flood or control the level of water in each distinct field. In some cases, the water is pumped, or lifted by human or animal power to the level of the land. The water application efficiency of surface irrigation is typically lower than other forms of irrigation.

Surface irrigation is even used to water landscapes in certain areas, for example, in and around Phoenix, Arizona. The irrigated area is surrounded by a berm and the water is delivered according to a schedule set by a local irrigation district.

Micro-irrigation

Micro-irrigation, sometimes called localized irrigation, low volume irrigation, or trickle irrigation is a system where water is distributed under low pressure through a piped network, in a pre-determined pattern, and applied as a small discharge to each plant or adjacent to it. Traditional drip irrigation using individual emitters, subsurface drip irrigation (SDI), micro-spray or micro-sprinkler irrigation, and mini-bubbler irrigation all belong to this category of irrigation methods.

Drip irrigation

Drip (or micro) irrigation, also known as trickle irrigation, functions as its name suggests. In this system water falls drop by drop just at the position of roots. Water is delivered at or near the root zone of plants, drop by drop. This method can be the most water-efficient method of irrigation, if managed properly, evaporation and runoff are minimized. The field water efficiency of drip irrigation is typically in the range of 80 to 90 percent when managed correctly. In modern agriculture, drip irrigation is often combined with plastic mulch, further reducing evaporation, and is also the means of delivery of fertilizer. The process is known as fertigation.

Deep percolation, where water moves below the root zone, can occur if a drip system is operated for too long or if the delivery rate is too high. Drip irrigation methods range from very high-tech and computerized to low-tech and labor-intensive. Lower water pressures are usually needed than for most other types of systems, with the exception of low energy center pivot systems and surface irrigation systems, and the system can be designed for uniformity throughout a field or for precise water delivery to individual plants in a landscape containing a mix of plant species. Although it is difficult to regulate pressure on steep slopes, pressure compensating emitters are available, so the field does not have to be level. High-tech solutions involve precisely calibrated emitters located along lines of tubing that extend from a computerized set of valves.

Sprinkler Irrigation

In sprinkler or overhead irrigation, water is piped to one or more central locations within the field and distributed by overhead high-pressure sprinklers or guns. A system utilizing sprinklers, sprays, or guns mounted overhead on permanently installed risers is often referred to as a solid-set irrigation system. Higher pressure sprinklers that rotate are called rotors and are driven by a ball drive, gear drive, or impact mechanism. Rotors can be designed to rotate in a full or partial circle. Guns are similar to rotors, except that they generally operate at very high pressures of 40 to 130 lbf/in^2 (275 to 900 kPa) and flows of 50 to 1200 US gal/min (3 to 76 L/s), usually with nozzle diameters in the range of 0.5 to 1.9 inches (10 to 50 mm). Guns are used not only for irrigation, but also for industrial applications such as dust suppression and logging.

Sprinklers can also be mounted on moving platforms connected to the water source by a hose. Automatically moving wheeled systems known as traveling sprinklers may irrigate areas such as small farms, sports fields, parks, pastures, and cemeteries unattended. Most of these utilize a length of polyethylene tubing wound on a steel drum. As the tubing is wound on the drum powered by the irrigation water or a small gas engine, the sprinkler is pulled across the field. When the sprinkler arrives back at the reel the system shuts off. This type of system is known to most people as a "waterreel" traveling irrigation sprinkler and they are used extensively for dust suppression, irrigation, and land application of waste water.

Other travelers use a flat rubber hose that is dragged along behind while the sprinkler platform is pulled by a cable.

Center pivot

Center pivot irrigation is a form of sprinkler irrigation utilising several segments of pipe (usually galvanized steel or aluminium) joined together and supported by trusses, mounted on wheeled towers with sprinklers positioned along its length. The system moves in a circular pattern and is fed with water from the pivot point at the center of the arc. These systems are found and used in all parts of the world and allow irrigation of all types of terrain. Newer systems have drop sprinkler heads as shown in the image that follows.

As of 2017 most center pivot systems have drops hanging from a U-shaped pipe attached at the top of the pipe with sprinkler heads that are positioned a few feet (at most) above the crop, thus limiting evaporative losses. Drops can also be used with drag hoses or bubblers that deposit the water directly on the ground between crops. Crops are often planted in a circle to conform to the center pivot. This type of system is known as LEPA (Low Energy Precision Application). Originally, most center pivots were water-powered. These were replaced by hydraulic systems (T-L Irrigation) and electric-motor-driven systems (Reinke, Valley, Zimmatic). Many modern pivots feature GPS devices.

Irrigation by Lateral move (side roll, wheel line, wheelmove)

A series of pipes, each with a wheel of about 1.5 m diameter permanently affixed to its midpoint, and sprinklers along its length, are coupled together. Water is supplied at one end using a large hose. After sufficient irrigation has been applied to one strip of the field, the hose is removed, the water drained from the system, and the assembly rolled either by hand or with a purpose-built mechanism, so that the sprinklers are moved to a different position across the field. The hose is reconnected. The process is repeated in a pattern until the whole field has been irrigated. This system is less expensive to install than a center pivot, but much more labor-intensive to operate - it does not travel automatically across the field: it applies water in a stationary strip, must be drained, and then rolled to a new strip. Most systems use 4 or 5-inch (130 mm) diameter aluminum pipe. The pipe doubles both as water transport and as an axle for rotating all the wheels. A drive system (often found near the centre of the wheel line) rotates the clamped-together pipe sections as a single axle, rolling the whole wheel line. Manual adjustment of individual wheel positions may be necessary if the system becomes misaligned.

Wheel line systems are limited in the amount of water they can carry, and limited in the height of crops that can be irrigated. One useful feature of a lateral move system is that it consists of sections that can be easily disconnected, adapting to field shape as the line is moved. They are most often used for small, rectilinear, or oddly-shaped fields, hilly or mountainous regions, or in regions where labor is inexpensive.

LAWN SPRINKLER SYSTEMS

A lawn sprinkler system is permanently installed, as opposed to a hose-end sprinkler, which is portable. Sprinkler systems are installed in residential lawns, in commercial landscapes, for churches and schools, in public parks and cemeteries, and on golf courses. Most of the components of these irrigation systems are hidden under ground, since aesthetics are important in a landscape. A typical lawn sprinkler system will consist of one or more zones, limited in size by the capacity of the water source. Each zone will cover a designated portion of the landscape. Sections of the landscape will usually be divided by microclimate, type of plant material, and type of irrigation equipment. A landscape irrigation system may also include zones containing drip irrigation, bubblers, or other types of equipment besides sprinklers.

Although manual systems are still used, most lawn sprinkler systems may be operated automatically using an irrigation controller, sometimes called a clock or timer. Most automatic systems employ electric solenoid valves. Each zone has one or more of these valves that are wired to the controller. When the controller sends power to the valve, the valve opens, allowing water to flow to the sprinklers in that zone.

There are two main types of sprinklers used in lawn irrigation, pop-up spray heads and rotors. Spray heads have a fixed spray pattern, while rotors have one or more streams that rotate. Spray heads are used to cover smaller areas, while rotors are used for larger areas. Golf course rotors are sometimes so large that a single sprinkler is combined with a valve and called a 'valve in head'. When used in a turf area, the sprinklers are installed with the top of the head flush with the ground surface. When the system is pressurized, the head will pop up out of the ground and water the desired area until the valve closes and shuts off that zone. Once there is no more pressure in the lateral line, the sprinkler head will retract back into the ground. In flower beds or shrub areas, sprinklers may be mounted on above ground risers or even taller pop-up sprinklers may be used and installed flush as in a lawn area.

Hose-end sprinklers

There are many types of hose-end sprinklers. Many of them are smaller versions of larger agricultural and landscape sprinklers, sized to work with a typical garden hose. Some have a spiked base allowing them to be temporarily stuck in the ground, while others have a sled base designed to be dragged while attached to the hose.

Sub-irrigation

Subirrigation has been used for many years in field crops in areas with high water tables. It is a method of artificially raising the water table to allow the soil to be moistened from below the plants' root zone. Often those systems are located on permanent grasslands in lowlands or river valleys and combined with drainage infrastructure. A system of pumping stations, canals, weirs and gates allows it to increase or decrease the water level in a network of ditches and thereby control the water table.

Subirrigation is also used in commercial greenhouse production, usually for potted plants. Water is delivered from below, absorbed upwards, and the excess collected for recycling. Typically, a solution of water and nutrients floods a container or flows through a trough for a short period of time, 10-20 minutes, and is then pumped back into a holding tank for reuse. Sub-irrigation in greenhouses requires fairly sophisticated, expensive equipment and management. Advantages are water and nutrient conservation, and labor savings through reduced system maintenance and automation. It is similar in principle and action to subsurface basin irrigation.

Another type of subirrigation is the self-watering container, also known as a sub-irrigated planter. This consists of a planter suspended over a reservoir with some type of wicking material such as a polyester rope. The water is drawn up the wick through capillary action.

Subsurface textile irrigation

Subsurface Textile Irrigation (SSTI) is a technology designed specifically for subirrigation in all soil textures from desert sands to heavy clays. A typical subsurface textile irrigation system has an impermeable base layer (usually polyethylene or polypropylene), a drip line running along that base, a layer of geotextile on top of the drip line and, finally, a narrow impermeable layer on top of the geotextile (see diagram). Unlike standard drip irrigation, the spacing of emitters in the drip pipe is not critical as the geotextile moves the water along the fabric up to 2 m from the dripper. The impermeable layer effectively creates an artificial water table.

Irrigation vs. Rain-Fed Agriculture

There are two main ways that farmers and ranchers use agricultural water to cultivate crops:

- Rain-fed farming
- Irrigation

Rain-fed farming is the natural application of water to the soil through direct rainfall. Relying on rainfall is less likely to result in contamination of food products but is open to water shortages when rainfall is reduced. On the other hand, artificial applications of water increase the risk of contamination.

Irrigation is the artificial application of water to the soil through various systems of tubes, pumps, and sprays. Irrigation is usually used in areas where rainfall is irregular or dry times or drought is expected. There are many types of irrigation systems, in which water is supplied to the entire field uniformly. Irrigation water can come from groundwater, through springs or wells, surface water, through rivers, lakes, or reservoirs, or even other sources, such as treated wastewater or desalinated water. As a result, it is critical that farmers protect their agricultural water source to minimize the potential for contamination. As with any groundwater removal, users of irrigation water need to be careful in not pumping groundwater out of an aquifer faster than it is being recharged.

Irrigation water use

Think of what your supper table might be like if water was not used to irrigate crops. Do you think you could survive very long without heaping servings of eggplant, beets, brussels sprouts, and rutabagas? Irrigation water is essential for keeping fruits, vegetables, and grains growing to feed the world's population, and this has been a constant for thousands of years.

The Texas Water Development Board (TWDB) is legislatively directed to plan for, and to assist financially, the development and management of the water resources of Texas. The usefulness of the TWDB's water planning and funding activities is dependent upon the accuracy and completeness of the information that water users provide. To this end, the TWDB annually collects and maintains accurate information concerning the current use of water.

TWDB staff is responsible for providing data to assist regional planning groups in planning for future water supply and demand, including these agricultural irrigation estimates. TWDB does not have the authority to require producers to report actual irrigation water use.

Agriculture, and especially irrigated agriculture, is the sector with by far the largest consumptive water use and water withdrawal. To estimate the pressure of irrigation on the available water resources, an assessment has to be made both of irrigation water requirement and of irrigation water withdrawal.

Irrigation water requirement depends on the crop water requirement and the water naturally available to the crops (effective precipitation, soil moisture, etc.). While part can be estimated based on climatic conditions, part results from physiological processes at plant level for which actual figures are not available, calling for modelling. In this review, the model's inputs are AQUASTAT data related to the corresponding irrigated crops: areas of harvested irrigated crops, cropping patterns and cropping intensity, converted into irrigated crop calendars.

Irrigation water withdrawal (or water withdrawal for irrigation) largely exceeds irrigation water requirement due to significant losses in distribution and application. Although available for some countries, figures of irrigation water withdrawal are easily confused with agricultural water withdrawal. Moreover, in the absence of direct measurement and due to the complexity of assessment methods, they are not always reliable. These difficulties explain that such figures are not always available at country level.

Nonetheless, a review of these variables at country level is necessary to improve the overall quality of global water resources monitoring. The main objective of this review is thus to update the AQUASTAT database with modelled irrigation water requirement, to supplement national data with estimates of irrigation water withdrawal and to check data accuracy and quality thanks to water requirement ratios comparing the two variables. In doing so, the database will provide policy- and decisions-makers as well as the scientific community with a complete dataset containing reliable data, calculated in a uniform way, and comparable with each other at country level.

The first review took place around 2000 within the framework of the preparation of FAO's global perspective study World agriculture: towards 2015/2030: an FAO perspective for 90 developing countries and countries in transition. The spatial coverage of the data of the present updated review consists of 165 countries(out of the total of 198 countries in 2012) and 2 territories, corresponding to those practicing irrigation and for which data on irrigated areas and on irrigated crops are available. Among those 167 countries and territories, 41 are high income countries and 1 high income territory, 89 are middle income countries and 1 middle income territory, and 35 are low income countries.

Irrigation water withdrawals for the Nation, 2010

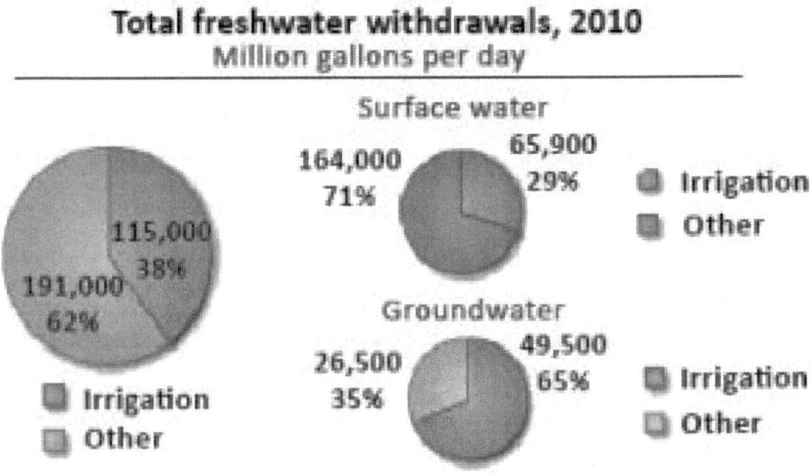

For 2010, total irrigation withdrawals were 115,000 Mgal/d, or 129,000 thousand acre-ft/yr, which accounted for 38 percent of total freshwater withdrawals and 61 percent of total freshwater withdrawals for all categories excluding thermoelectric power. Total irrigation withdrawals were 9 percent less than in 2005. Withdrawals from surface-water sources were 65,900 Mgal/d, which accounted for 57 percent of the total irrigation withdrawals, and were almost 12 percent less than in 2005. Groundwater withdrawals for 2010 were 49,500 Mgal/d, or 6 percent less than in 2005.

About 62,400 thousand acres were irrigated in 2010, an increase of about 950 thousand acres (1.5 percent) from 2005. About 31,600 thousand acres (51 percent) were irrigated with sprinkler systems, 26,200 thousand acres with surface (flood), and 4,610 thousand acres with microirrigation systems. The national average application rate for 2010 was 2.07 acrefeet per acre, or 11 percent less than the 2005 average of 2.32 acre-feet per acre.

Irrigation water withdrawals

The majority of total U.S. irrigation withdrawals (83 percent) and irrigated acres (74 percent) were in the 17 conterminous Western States (west of solid line in figure 7), which are typical of areas where average annual precipitation is less than 20 inches and generally insufficient to support crops without supplemental water. Surface water was the primary source of water in the arid West, except in Kansas, Oklahoma, Nebraska, Texas, and South Dakota, where more groundwater was used. The 17 Western States cumulatively accounted for 93 percent of total surface-water irrigation withdrawals and 69 percent of total groundwater irrigation withdrawals.

Because the 17 Western States accounted for the majority of total irrigation withdrawals, changes in those States had a great effect on the overall total. Total irrigation withdrawals declined noticeably in Nebraska, Montana, Idaho, Colorado, and California. Groundwater irrigation withdrawals declined in the West and increased in the East, and surface-water irrigation withdrawals declined in both regions. Total irrigated acres increased in both regions --1 percent (568 thousand acres) in the West, and 2 percent (381 thousand acres) in the East. In the West, the total number of acres irrigated by the less-efficient surface-irrigation methods decreased by about 500 thousand acres, and the number of acres irrigated by more efficient sprinkler (including microirrigation) methods increased by about 1,080 thousand acres.

Trends in irrigation water withdrawals, 1950-2010

Since 1950, irrigation has represented about 64 percent of total withdrawals, excluding those for thermoelectric power. Withdrawals for irrigation increased by more than 68 percent from 1950 to 1980 (from 89,000 to 150,000 Mgal/d). Withdrawals have decreased since 1980 and have stabilized at between 135,000 and 139,000 Mgal/d between 1985 and 2000, and 115,000 in 2010. Depending on the geographic area of the United States, this overall decrease can be attributed to climate, crop type, advances in irrigation efficiency, and higher energy costs.

Surface water historically has been the primary source for irrigation, although data show an increasing usage of groundwater since 1950. During 1950, 77 percent of all irrigation withdrawals were surface water, most of which was used in the western States. By 2010, surface-water withdrawals comprised only 57 percent of the total. Groundwater withdrawals for irrigation during 2010 were more than double withdrawals in 1950. Most of this increase occurred from 1965 through 1980.

The trend in irrigation withdrawals is seen as the green bar in the chart below

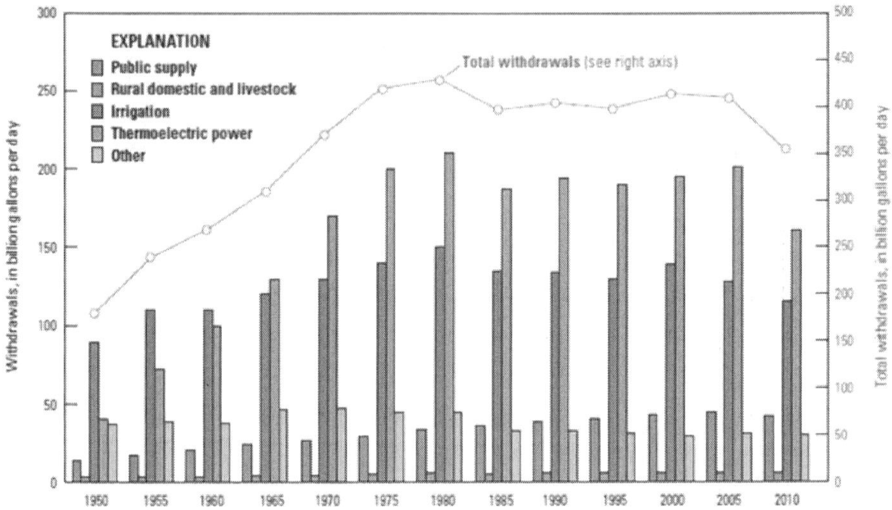

SCARCITY OF WATER IN AGRICULTURE

50 years ago, the common perception was that water was an infinite resource. At that time, there were fewer than half the current number of people on the planet. Affluence was not as high, individuals consumed fewer calories and ate less meat, so less water was needed to produce their food. They required a third of the volume of water we presently take from rivers. Today, the competition for water resources is much more intense. This is because there are now nearly eight billion people on the planet, their consumption of meat and vegetables is rising. Competition for water from industry, urbanisation and biofuel crops is rising congruently. To avoid a global water crisis, farmers will have to make strides to increase productivity to meet growing demands for food, while industry and cities find ways to use water more efficiently.

Successful agriculture is dependent upon farmers having sufficient access to water. However, water scarcity is already a critical constraint to farming in many parts of the world. Physical water scarcity is where there is not enough water to meet all demands, including that needed for ecosystems to function effectively. Arid regions frequently suffer from physical water scarcity. It also occurs where water seems abundant but where resources are over-committed. This can happen where there is over development of hydraulic infrastructure, usually for irrigation. Symptoms of physical water scarcity include environmental degradation and declining groundwater. Economic scarcity, meanwhile, is caused by a lack of investment in water or insufficient human capacity to satisfy the demand for water.

Symptoms of economic water scarcity include a lack of infrastructure, with people often having to fetch water from rivers for domestic and agricultural uses. Some 2.8 billion people currently live in water-scarce areas.

SUSTAINABLE WATER USE

While it is evident that water use affects environmental degradation and economic growth, it is also sparking innovation regarding new irrigation methods. In 2006, the USDA predicted that if the agricultural sector improved water efficiency by just 10 percent farms could annually save upwards of $200 million per year. Many of the practices which cut water use are cost effective. Farmers who use straw, compost, or mulch around their crops will reduce evaporation by about 75 percent. This would also reduce the number of weeds and save a farmer from using herbicides. Mulches or ground covers also allow the soils to absorb more water by reducing compaction. The use of white or pale gravel is also practiced, as it reduces evaporation and keeps soil temperatures low by reflecting sunlight.

In addition to reducing water loss at the sink, there are also more sustainable ways to harvest water. Many modern small (non-industrial) farmers are using rain barrels to collect the water needed for their crops and livestock. On average, rainwater harvesting reduces the cost of water per cubic meter in half. This would also greatly reduce the stress on local aquifers and wells. Because farmers use the roofs of their buildings to gather this water, this also reduced rainwater runoff and soil erosion on and around their farm.

Water resources engineering

Water resources engineering is the study and management of equipment, facilities and techniques that are used to manage and preserve life's most plentiful resource. In addition to assessing how and the best ways in which to control water as it pertains to water-related activities – such as irrigation, waste disposal and canal development – water resource engineers are also frequently involved in water management to ensure that it's safe to drink both for humans, plants and animal usage. As previously referenced, surface water makes up about 71% of the planet, which is the equivalent of roughly 326 million cubic miles. At the same time, though, just 3% of the Earth's water is fresh, according to the Bureau of Reclamation. And of this total, 2.5% of it is out of reach, contained in the soil, polar ice caps, the atmosphere and glaciers or too polluted to use safely.

Water resource engineers may be tasked with the awesome responsibility of ensuring that the planning and management of available water supply are adequately leveraged and remain safe to use for as long as possible. They may also be involved in water treatment so that the quality of water is improved upon for various end uses, whether that's recreationally, commercially or industrially.

Water resources engineering important

Resources, by their very nature, are finite. There are only a small handful that are naturally renewable – such as wind, solar, hydro and biomass. While water may be renewable in terms of the many different ways it can be used and reused, it's not as abundant as it once was, which many earth scientists and climatologists point to as a function of climate change.

The Bureau of Reclamation provides some perspective as to just how limited this resource is in terms of usability, despite its vastness. If the world's water supply were roughly 26 gallons, the amount of freshwater available for safe usage would be the equivalent to 0.003 liters. That's equal to roughly a half-teaspoon.

Water resource engineers may be charged with developing new systems or processes for private or government entities that can preserve freshwater sources and find new ones. This may require the assistance of civil engineers involved as well, designing water purification methods through desalination or creating new equipment for contaminant transport when water is used for irrigation purposes. Understanding what works and what doesn't when it comes to water resource management is often a combined effort and may involve a number of different analyses, including hydrologic, which is the study of the water cycle and directions in which it flows, which may be influenced by weather and other environmental forces.

Like most other professions, the average water engineer salary is largely a function of how much experience they have and their level of education. At a bare minimum, most water resource engineering jobs require a bachelor's degree. The more experience you have, such as a master's degree in environmental engineering, the more you stand to earn. According to the most recent figures available from the Bureau of Labor Statistics, professionals in environmental engineering typically make approximately $87,600, which was the median in 2018. However, the top 10% earned nearly $137,100 and were generally employed by the federal government.

9

TYPES OF DAMS

A dam is a barrier that impounds water or underground streams. Reservoirs created by dams not only suppress floods but also provide water for activities such as irrigation, human consumption, industrial use, aquaculture, and navigability. Hydropower is often used in conjunction with dams to generate electricity. A dam can also be used to collect water or for storage of water which can be evenly distributed between locations. Dams generally serve the primary purpose of retaining water, while other structures such as floodgates or levees (also known as dikes) are used to manage or prevent water flow into specific land regions. For details see: Purposes of Dams

Dams could be of temporary or permanent nature. Dams may be built by constructing an embankment across the river at some suitable location.

A dam is a hydraulic structure of fairly impervious material built across a river to create a reservoir on its upstream side for impounding water for various purposes. These purposes may be Irrigation, Hydropower, Water-supply, Flood Control, Navigation, Fishing and Recreation. Dams may be built to meet the one of the above purposes or they may be constructed fulfilling more than one. As such, Dam can be classified as: Single-purpose and Multipurpose Dam.

DIFFERENT PARTS & TERMINOLOGIES OF DAMS

Crest: The top of the Dam. These may in some cases be used for providing a roadway or walkway over the dam.

Parapet walls: Low Protective walls on either side of the roadway or walkway on the crest.

Heel: Portion of Dam in contact with ground or river-bed at upstream side.

Toe: Portion of dam in contact with ground or river-bed at downstream side.

Spillway: It is the arrangement made (kind of passage) near the top of dam for the passage of surplus/ excessive water from the reservoir.

Abutments: The valley slopes on either side of the dam wall to which the left & right end of dam are fixed to.

Gallery: Level or gently sloping tunnel like passage (small room like space) at transverse or longitudinal within the dam with drain on floor for seepage water. These are generally provided for having space for drilling grout holes and drainage holes. These may also be used to accommodate the instrumentation for studying the performance of dam.

Sluice way: Opening in the dam near the base, provided to clear the silt accumulation in the reservoir.

Free board: The space between the highest level of water in the reservoir and the top of the dam.

Dead Storage level: Level of permanent storage below which the water will not be withdrawn.

Diversion Tunnel: Tunnel constructed to divert or change the direction of water to bypass the dam construction site. The dam is built while the river flows through the diversion tunnel.

VARIOUS TYPES OF DAMS

Dams can be classified in number of ways. But most usual ways of classification of dams are mentioned below:

Based on the functions of dam, it can be classified as follows:

Storage dams: They are constructed to store water during the rainy season when there is a large flow in the river. Many small dams impound the spring runoff for later use in dry summers. Storage dams may also provide a water supply, or improved habitat for fish and wildlife. They may store water for hydroelectric power generation, irrigation or for a flood control project. Storage dams are the most common type of dams and in general the dam means a storage dam unless qualified otherwise.

Diversion dams: A diversion dam is constructed for the purpose of diverting water of the river into an off-taking canal (or a conduit). They provide sufficient pressure for pushing water into ditches, canals, or other conveyance systems. Such shorter dams are used for irrigation, and for diversion from a stream to a distant storage reservoir. A diversion dam is usually of low height and has a small storage reservoir on its upstream. The diversion dam is a sort of storage weir which also diverts water and has a small storage. Sometimes, the terms weirs and diversion dams are used synonymously.

Detention dams: Detention dams are constructed for flood control. A detention dam retards the flow in the river on its downstream during floods by storing some flood water. Thus the effect of sudden floods is reduced to some extent. The water retained in the reservoir is later released gradually at a controlled rate according to the carrying capacity of the channel downstream of the detention dam. Thus the area downstream of the dam is protected against flood.

Debris dams: A debris dam is constructed to retain debris such as sand, gravel, and drift wood flowing in the river with water. The water after passing over a debris dam is relatively clear.

Coffer dams: It is an enclosure constructed around the construction site to exclude water so that the construction can be done in dry. A cofferdam is thus a temporary dam constructed for facilitating construction. A coffer dam is usually constructed on the upstream of the main dam to divert water into a diversion tunnel (or channel) during the construction of the dam. When the flow in the river during construction of the dam is not much, the site is usually enclosed by the coffer dam and pumped dry. Sometimes a coffer dam on the downstream of the dam is also required.

Based on structure and design, dams can be classified as follows:

Gravity dam: The gravity dam is a massive structure made of masonry or concrete with earth and rock fill. Gravity dams use their own weight to resist opposing forces and as such require a hard bedrock foundation. They rely on their great weight and size for stability. The gravity dam is the most commonly built dam in the world. One of the reasons for this is the simplicity of design. Unfortunately it requires great quantities of material to construct.

A gravity dam is a massive sized dam fabricated from concrete or stone masonry. They are designed to hold back large volumes of water. By using concrete, the weight of the dam is actually able to resist the horizontal thrust of water pushing against it. This is why it is called a gravity dam. Gravity essentially holds the dam down to the ground, stopping water from toppling it over.

Gravity dams are well suited for blocking rivers in wide valleys or narrow gorge ways. Since gravity dams must rely on their own weight to hold back water, it is necessary that they are built on a solid foundation of bedrock.

Examples of Gravity dam: Grand Coulee Dam (USA), (Nagarjuna Sagar Dam (India) and Itaipu Dam (Between Brazil and Paraguay).

Earth Dams: An earth dam is made of earth (or soil) built up by compacting successive layers of earth, using the most impervious materials to form a core and placing more permeable substances on the upstream and downstream sides. A facing of crushed stone prevents erosion by wind or rain, and an ample spillway, usually of concrete, protects against catastrophic washout should the water overtop the dam. Earth dam resists the forces exerted upon it mainly due to shear strength of the soil.

Although the weight of the earth dam also helps in resisting the forces, the structural behavior of an earth dam is entirely different from that of a gravity dam. The earth dams are usually built in wide valleys having flat slopes at flanks (abutments).The foundation requirements are less stringent than those of gravity dams, and hence they can be built at the sites where the foundations are less strong. They can be built on all types of foundations. However, the height of the dam will depend upon the strength of the foundation material.

Examples of earthfill dam: Rongunsky dam (Russia) and New Cornelia Dam (USA).

Rockfill Dams: A rockfill dam is built of rock fragments and boulders of large size. An impervious membrane is placed on the rockfill on the upstream side to reduce the seepage through the dam. The membrane is usually made of cement concrete or asphaltic concrete. In early rockfill dams, steel and timber membrane were also used, but now they are obsolete.

A dry rubble cushion is placed between the rockfill and the membrane for the distribution of water load and for providing a support to the membrane. Sometimes, the rockfill dams have an impervious earth core in the middle to check the seepage instead of an impervious upstream membrane. The earth core is placed against a dumped rockfill. It is necessary to provide adequate filters between the earth core and the rockfill on the upstream and downstream sides of the core so that the soil particles are not carried by water and piping does not occur. The side slopes of rockfill are usually kept equal to the angle of repose of rock, which is usually taken as 1.4:1 (or 1.3:1). Rockfill dams require foundation stronger than those for earth dams.

Examples of rockfill dam: Mica Dam (Canada) and Chicoasen Dam (Mexico)

Arch Dams

An arch dam is curved in plan, with its convexity towards the upstream side. An arch dam transfers the water pressure and other forces mainly to the abutments by arch action. An arch dam is quite suitable for narrow canyons with strong flanks which are capable of resisting the thrust produced by the arch action.

The section of an arch dam is approximately triangular like a gravity dam but the section is comparatively thinner. The arch dam may have a single curvature or double curvature in the vertical plane. Generally, the arch dams of double curvature are more economical and are used in practice.

An arch dam is a type of dam that is curved and commonly built with concrete. The arch dam is a structure that is designed to curve upstream so that the force of the water against it, known as hydrostatic pressure, presses against the arch, compressing and strengthening the structure as it pushes into its foundation or abutments. An arch dam is most suitable for narrow gorges or canyons with steep walls of stable rock to support the structure and stresses.

Since they are thinner than any other dam type, they require much less construction material, making them economical and practical in remote areas. The development of arch dams throughout history began with the Romans in the 1st century BC and after several designs and techniques were developed, relative uniformity was achieved in the 20th century.

The first known arch dam, the Glanum Dam, also known as the Vallon de Baume Dam, was built by the Romans in France and it dates back to the 1st century BC. The dam was about 12m high and 18m in length. Its radius was about 14m and it consisted of two masonry walls. The Romans built it to supply nearby Glanum with water.

The Monte Novo Dam in Portugal was another early arch dam built by the Romans in 300 AD. It was 5.7m high and 52m long with a radius of 19m. The curved ends of the dam met with two winged walls that were later supported by two buttresses. The dam also contained two water outlets to drive mills downstream.

The Dara Dam was another arch dam built by the Romans in which the historian Procopius would write of its design: "This barrier was not built in a straight line, but was bent into the shape of a crescent, so that the curve, by lying against the current of the river, might be able to offer still more resistance to the force of the stream."

The Mongolians also built arch dams in modern-day Iran. Their earliest was the Kebar Dam built around 1300 which was 26m high, 55m long and had a radius of 35m. Their second dam was built around 1350 and is called the Kurit Dam. After 4m was added to the dam in 1850, it became 64m tall and remained the tallest dam in the world until the early 20th century. The Kurit Dam was of masonry design and built in a very narrow canyon. The canyon was so narrow that its crest length is only 44% of its height. The dam is still erect even though part of its lower downstream face fell off.

The Elche Dam in Elche, Spain was a post-medieval arch dam built in the 1630s by Joanes del Temple and the first in Europe since the Romans. The dam was 26m high, 75m long and had a radius of 62m. This arch dam also rest on winged walls that served as abutments.

In the 20th century, the world's first variable radius arch dam was built on the Salmon Creek near Juneau, Alaska. The Salmon Creek Dam's upstream face bulged upstream which relieved pressure on the stronger curved lower arches near the abutments.

The dam also had a larger toe which off-set pressure on the upstream heel of the dam which now curved more downstream. The technology and economical benefits of the Salmon Creek Dam allowed for larger and taller dam designs. The dam was therefore revolutionary and similar designs were soon adopted around the world particularly by the U.S. Bureau of Reclamation.

Arch dam designs would continue to test new limits and designs such as the double and multiple-curve. The Swiss engineer Alfred Stucky and the U.S. Bureau of Reclamation would develop a method of weight and stress distribution in the 1960s and arch dam construction in the United States would see its last surge then with dams like the 143m double-curved Morrow Point Dam in Colorado. By the late 20th century, arch dam design reached a relative uniformity in design around the world.

Currently, the tallest and largest arch dam in the world is the 272 m Inguri Dam in Georgia which was completed in 1980. The longest multiple arch with buttress dam in the world is the Daniel-Johnson Dam in Québec, Canada. It is 214 m high and 1,314 meters long across its crest. It was completed in 1968 and put in service in 1970. China is currently constructing the Jinping 1 Dam in Sichuan which when complete will be world's tallest arch dam at 305m.

Design of Arch Dams

Arch dams require a high level of stress and force analysis in order to create a sufficient design. The main force against an arch dam is the hydrostatic pressure provided by the reservoir behind it, uplift which is water pressure beneath the dam, the weight of the dam itself and finally all the forces combined. Other forces that affect a dam include, but are not limited to, temperature, chemical reactions, settling, silt accumulation and earthquakes.

Most often, the arch dam is made of concrete and placed in a "V" shaped valley. The foundation or abutments for an arch dam must be very stable and proportionate to the concrete. There are two basic designs for an arch dam: constant radius which have constant radius of curvature and variable radius dams which have both upstream and downstream curves that systematically decrease in radius below the crest.

A dam that is double-curved in both its horizontal and vertical planes may be called a dome dam. Arch dams with more than one contiguous arch or plane are described as multiple-arch dams. Early examples include the Roman Esparragalejo Dam and Muro Dam with later examples such as the Daniel-Johnson Dam (1968) and Itaipu Dam (1982). At an increased cost, an arch dam can have steel rods or pre-stressed steel cable reinforcements and therefore require less concrete than does a gravity dam or arch-gravity dam. As arch dams are commonly constructed in remote mountainous areas, less material is beneficial when it has to be delivered with difficulty to the site. However, as a result of the failure of the Gleno Dam shortly after it was constructed in 1923, the construction of new multiple arch dams has become less popular.

Arch dams are generally classified as thin, medium or thick. If width of an arch dam's base is 2/10 of the dam's height or less it is thin. Any base and height ratio greater than 2/10 but less than 3/10 is considered medium. Thick is a ratio of 3/10 or greater. Additionally, an arch dam's base is historically just as thick as the crest but often it is twice as thick as the crest. Contraction joints are normally placed every 20 m in the arch dam and are later filled with grout after the control cools and cures.

Examples of arch dams

- Inguri Dam
- Buchanan Dam (example of Multiple-Arch type)
- Glen Canyon Dam
- Hoover Dam
- Idukki Dam
- Karun-3 dam
- Daniel-Johnson Dam

Buttress Dams: Buttress dams are of three types : (i) Deck type, (ii) Multiple-arch type, and (iii) Massive-head type. A deck type buttress dam consists of a sloping deck supported by buttresses. Buttresses are triangular concrete walls which transmit the water pressure from the deck slab to the foundation. Buttresses are compression members.

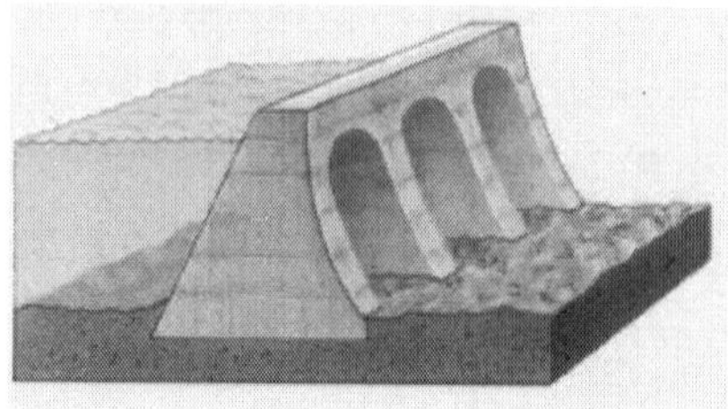

Buttresses are typically spaced across the dam site every 6 to 30 metre, depending upon the size and design of the dam. Buttress dams are sometimes called hollow dams because the buttresses do not form a solid wall stretching across a river valley. The deck is usually a reinforced concrete slab supported between the buttresses, which are usually equally spaced.

In a multiple-arch type buttress dam the deck slab is replaced by horizontal arches supported by buttresses. The arches are usually of small span and made of concrete. In a massive-head type buttress dam, there is no deck slab. Instead of the deck, the upstream edges of the buttresses are flared to form massive heads which span the distance between the buttresses. The buttress dams require less concrete than gravity dams. But they are not necessarily cheaper than the gravity dams because of extra cost of form work, reinforcement and more skilled labor. The foundation requirements of a buttress dam are usually less stringent than those in a gravity dam.

Examples of Buttress Dam: Bartlett dam (USA) and The Daniel-Johnson Dam (Canada)

Steel Dams: A steel dam consists of a steel framework, with a steel skin plate on its upstream face. Steel dams are generally of two types: (i) Direct-strutted steel dams, and (ii)

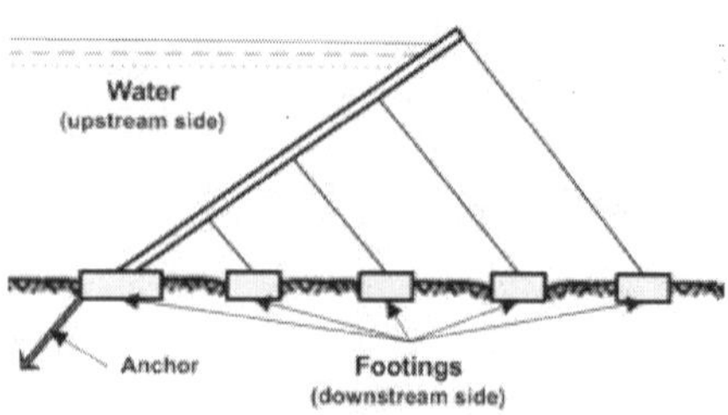

Cantilever type steel dams. In a direct strutted steel dam, the water pressure is transmitted directly to the foundation through inclined struts. In a cantilever type steel dam, there is a bent supporting the upper part of the deck, which is formed into a cantilever truss. This arrangement introduces a tensile force in the deck girder which can be taken care of by anchoring it into the foundation at the upstream toe. Hovey suggested that tension at the upstream toe may be reduced by flattening the slopes of the lower struts in the bent. However, it would require heavier sections for struts. Another alternative to reduce tension is to frame together the entire bent rigidly so that the moment due to the weight of the water on the lower part of the deck is utilised to offset the moment induced in the cantilever.

This arrangement would, however, require bracing and this will increase the cost. These are quite costly and are subjected to corrosion. These dams are almost obsolete. Steel dams are sometimes used as temporary coffer dams during the construction of the permanent dams. Steel coffer dams are supplemented with timber or earthfill on the inner side to make them water tight. The area between the coffer dams is dewatered so that the construction may be done in dry for the permanent dam.

Examples of Steel Dam: Redridge Steel Dam (USA) and Ashfork-Bainbridge Steel Dam (USA)

Timber Dams: Main load-carrying structural elements of timber dam are made of wood, primarily coniferous varieties such as pine and fir. Timber dams are made for small heads (2-4 m or, rarely, 4-8 m) and usually have sluices; according to the design of the apron they are divided into pile, crib, pile-crib, and buttressed dams.

The openings of timber dams are restricted by abutments; where the sluice is very long it is divided into several openings by intermediate supports: piers, buttresses, and posts. The openings are covered by wooden shields, usually several in a row one above the other. Simple hoists—permanent or mobile winches—are used to raise and lower the shields.

Nagarjuna Sagar Dam

Nagarjuna Sagar Dam is the world's largest masonry dam built across Krishna River in Nagarjuna Sagar, Nalgonda District of Andhra Pradesh, India. It is downstream to the Nagarjuna Sagar reservoir with a capacity of up to 11,472 million cubic metres which is the world's largest man-made lake with a concrete wall of that measures 6 ft (1.8 m). thick. The dam is 490 ft (150 m). tall and 16 km long with 26 gates which are 42 ft (13 m). wide and 45 ft (14 m). tall.

It is one of the earliest irrigation and hydro-electric projects in India. The dam provides irrigation water to the Nalgonda District, Prakasam District, Khammam District and Guntur District.

The proposal to construct a dam to use the excess waters of the Krishna river was put forward by the British rulers in 1903. Siddeswaram, Hyderabad and Pulichintala were identified as the suitable locations for the reservoirs. The perseverance of the Raja of Muktyala paved way for the site identification, design and construction of the dam.

The dam water was released by the then Prime Minister's daughter, Indira Gandhi in 1967. The construction of the dam submerged an ancient Buddhist settlement, Nagarjunakonda, which was the capital of the Ikshvaku dynasty in the 1st and 2nd centuries, the successors of the Satavahanas in the Eastern Deccan.

Excavations here had yielded 30 Buddhist monasteries, as well as art works and inscriptions of great historical importance. In advance of the reservoir's flooding, monuments were dug up and relocated. Some were moved to Nagarjuna's Hill, now an island in the middle of the reservoir. Others were moved to the mainland.

Effect of the project

The project benefited farmers in the districts of Guntur, Prakasam, Krishna, Nalgonda and Khammam. The right canal (a.k.a Jawahar canal) is 203 km long and irrigates 1.113 million acres (4,500 km^2) of land. The left canal (a.k.a Lalbahadur Shastri canal) is 295 km long and irrigates 0.32 million acres (800 km^2) of land in Nalgonda and Khammam districts of Telangana region. The project transformed the economy of above districts. 52 villages were submersed in water and 24000 people were affected. The relocation of the people was completed by 2007.

Power Generation

The hydroelectric plant has a power generation capacity of 815.6 MW with 8 units (1x110 MW+7x100.8 MW). First unit was commissioned on 7 March 1978 and 8th unit on 24 December 1985. The right canal plant has a power generation capacity of 90 MW with 3 units of 30 MW each. The left canal plant has a power generation capacity of 60 MW with 2 units of 30 MW each.

The dam is constructed on the border of Guntur and Nalgonda districts. The dam also provides drinking water to the Nalgonda town.

Statistics

- Catchment Area : 215000 km^2 (83012 sq mi)
- Masonry dam
  - Spillway of dam : 471 m
  - Non-over flow dam : 979 m
  - Length of Masonry dam : 1450 m
  - Maximum height : 125 m
- Earth dam
  - Total Length of Earth dam : 3414 m
  - Maximum height : 28 m

Daniel-Johnson Dam

The Daniel-Johnson Dam, formerly known as Manic-5, is a multiple arch buttress dam on the Manicouagan River which creates Manicouagan Reservoir. The dam is composed of 14 buttresses and 13 arches and is 214 km (133 mi) north of Baie-Comeau in Quebec, Canada. The dam was constructed between 1959 and 1970 for the purpose of hydroelectric power production and supplies water to the Manic-5 and Manic-5-PA power houses. The dam is 214 m (702 ft) tall, 1,314 m (4,311 ft) long and contains 2,200,000 m3 (2,880,000 cu yd) of concrete, making it the largest dam of its type in the world. The dam was named after Daniel Johnson, Sr., the 20th Premier of Quebec who was responsible for starting the project. Johnson died on 26 September 1968, on the day he was to preside over the scheduled inauguration of the dam. The facility is owned and operated by Hydro-Québec.

During the summers of 1919 and 1920, hydrological studies were conducted on the Manicouagan and Outardes rivers, flowing into the St. Lawrence near the town of Baie-Comeau. The combined flow was then estimated at 40 million cubic meters, making it one of the largest hydrological systems in Canada. Although harnessing of this potential was considered interesting, its distance from major load centers and the lack of roads in the area were identified as major drawbacks. In addition, building dams in the wilderness was considered too costly.

After the Second World War, the discovery of large iron deposits on the North Shore and the increased forestry activity led to a rapid development of the region. The area's largest cities, Sept-Iles and Baie-Comeau, were now linked to the rest of the province by a road. At the same time, industrial development in southern Quebec required a larger electric supply. Improvements in long distance electric transmission technologies were also a consideration at the time. Significant advances in the field, including the construction of two 315-kilovolt lines, between the Bersimis complex, west of the Manicouagan system, and Montreal, (completed in 1956), lifted another obstacle.

In 1955, Hydro-Québec launched an 5-year extensive assessment of the Manicouagan's suitability. These studies demonstrated the exceptional potential of the river and stressed the advantages of building a multi-dam system in order take full advantage of the terrain and water flows. The data collected at the time was so promising that Hydro-Québec did not wait for the working group to submit their final recommendations.

By 1959, a decision was made and construction of a 210 kilometres (130 mi) access road from Baie-Comeau was started. The original project, the Manicouagan-Outardes project, included the construction of five dams on the Manicouagan River (Manic-1, Manic-2, Manic-3, Manic-4 and Manic-5) and three on the Outardes River (Outardes-2, Outardes-3 and Outardes-4). However, a miscalculation prevented the construction of the Manic-4 dam and powerhouse because engineers realized early on that it would encroach on the Manic-3 reservoir.

In 1959, the Premier of Quebec, Maurice Duplessis, wanted a U.S. firm to construct the dam but Quebec's Minister of Water Resources at the time, Daniel Johnson was against the idea. Johnson believed that the skill of a Canadian firm and workers was well suited for the construction of such a complex dam. Duplessis' successor in 1959, Paul Sauvé agreed with Johnson and he continued to pave the way for the dam. Eventually, engineers would choose André Coyne's design for a multiple arch buttress dam as the most suitable and economical.

Construction

Construction on the dam's support facilities began in 1959 and the river diversion tunnels and foundation preparation commenced in 1960. To divert the Manicoagan River, workers blasted and dug two 2,000 ft (610 m) long and 45 ft (14 m) diameter tunnels through the gorge's solid granite west wall.

Workers used "jumbo" drilling platforms and the tunnels progressed 14 ft (4.3 m) each shift. To facilitate the river's diversion, two cofferdams were built. The first was a concrete arch upstream from the dam site which would block the river, forcing it into the diversion tunnels. The second cofferdam was downstream and prevented water from flowing back into the construction site. Each cofferdam was built on alluvial deposits (loose soil) so they needed to be watertight which was accomplished with either a grout curtain or deep supporting piles.

Once the river was diverted, workers pumped the remaining water out between the two cofferdams in order to prepare the site for construction. Once drained, workers excavated the alluvial deposits between the cofferdams creating a long pit that was 150 ft (46 m) deep at the center. This pit was then filled with concrete in the summer of 1962. To prevent seepage in the dam's foundation, a grout curtain was injected in the bedrock and a drainage network with 2,400 ft (730 m) of tunnels was constructed just downstream of the dam to collect water that may seep through.

After the extensive preparatory works were complete, the first concrete for the dam was poured on October 3, 1962. Concrete was poured day and night but was halted during the winter because of freezing temperatures. To organize the pouring, the dam was split into 45 ft (14 m) plots and each was raised about 5 or 6 ft. at a time. Workers had about 150 days - before seasonal flooding - to construct the dam to a height of at least 250 ft (76 m). Before flooding began, engineers planned to seal the diversion tunnels and begin filling the reservoir. To speed the pouring process, concrete was poured in casts by buckets that moved along three cableways suspended above the construction site. The deadline was met within the 150 days and a total of 1,000,000 cu yd (760,000 m3) of concrete was poured. The dam was eventually completed in 1968.

On September 25, 1968, the government-owned utility organized a ceremony to mark the completion of the Manicouagan-5 dam. Hundreds of dignitaries, politicians, utility executives, financiers, engineers and journalists were ferried by plane from Montreal, Quebec City and New York to the work-site to attend a banquet and a plaque unveiling ceremony.

Among the guests were Quebec Union Nationale Premier, Daniel Johnson Sr., his predecessor, Jean Lesage, and René Lévesque, the former Hydraulic Resources minister responsible for the consolidation of all investor-owned utilities into Hydro-Québec. Photographs taken at the banquet show the three men were in excellent spirits, holding hands and smiling, although relations between the Liberal leader and his former cabinet minister were strained by Lévesque's recent defection to the Mouvement Souveraineté-Association, a precursor of the Parti Québécois.

In his authorized biography, Hydro-Québec executive, Robert A. Boyd, recalls being woken up at 6 a.m. the next morning by his boss, Roland Giroux. "I've got bad news, Robert...", said Giroux, adding that he just found the Premier lying dead in his bed. Johnson's sudden demise sent shock waves at the work-site and across the province and the dedication ceremony was quickly cancelled.

On September 26, 1969, a year to the day after Johnson's death, the new Premier Jean-Jacques Bertrand accompanied by Johnson's widow and children, unveiled two plaques and officially dedicated the dam after his predecessor. Both plaques now sit side by side at the top of the complex.

Facilities

Dam

The design was chosen for strength and economical reasons as it used less concrete or material than a gravity or embankment dam. The Daniel-Johnson Dam is a 214 m (702 ft) tall and 1,312 m (4,304 ft) long multiple-arch buttress dam. Of the dam's 14 total buttresses, the two that form the center arch are 530 ft (160 m) apart at their base while the others are about 250 ft (76 m) apart. At its thickest point, the center, the dam is 22.5 m (74 ft) wide while the crest can reach about 3 m (10 ft) wide. Pressure from water behind the dam is transferred from the dam's arches to its buttresses and lastly into the ground or its foundation.

The dam was constructed with a high-quality concrete designed to withstand constant thawing and freezing associated with its environment. To further help the structure cope with the climate, engineers placed one inch steel reinforcing bars within the upstream and downstream faces of the dam. The concrete's strength in compression was initially 4,500 lbs. per square inch to meet an estimated 1,500 PSI within the structure. The upstream face of the dam was also coated in asphalt for protection against water. Despite the strength of the dam's concrete, two parallel inclined cracks were discovered on one of the arches shortly after construction.

Reservoir

The dam impounded the Manicougan River, which filled the Earth's fifth largest confirmed impact crater, the Manicouagan crater, creating Manicouagan Reservoir. Coincidentally, the reservoir itself is the fifth largest in the world as well. The reservoir has a maximum depth of 350 m (1,150 ft) feet, mean depth of 85 m (279 ft) and contains 142 km3 (34 cu mi) of water. While draining an area of 29,241 km (18,170 mi), it has a surface area of 1,950 km2 (750 sq mi) and shoreline of 1,322 km (821 mi).

The reservoir is a well known area for Atlantic salmon, lake trout and northern pike fishing, although tall trees flooded during the impoundment have not decomposed due to a lack of oxygen, which can sometimes interfere with the sport.

The priming of the reservoir also created a large artificial island in the center of the Manicouagan reservoir by merging two crescent-shaped lakes: Mouchalagane Lake on the western side and Manicouagan Lake on the eastern side. Covering an area of 2,020 km^2 (780 sq mi), René-Levasseur Island is considered to be the second largest island in the world located in a lake, in terms of area (the largest is Manitoulin Island in Lake Huron).

The island was named after René Levasseur, the chief engineer responsible for the construction of the Daniel-Johnson dam. Levasseur died at the age of 35, only days before the dam's inauguration.

Powerhouses

The dam fuels two powerhouses, the Manic-5 and Manic-5-PA. The first powerhouse consists of eight Francis turbines, capable of producing up to 1,528 MW of power, which went online in 1970. The second powerhouse, the Manic-5-PA (PA stands for puissance additionnelle or additional power), was commissioned in 1989, and consists of four Francis turbines of 1,064 MW in total installed capacity.

The designers of the Manic-5 decided on an above ground power house that was downstream of the dam for safety and cost. The intake was built on the east side of the dam and supplies two 3,400 ft (1,000 m), long 36 ft (11 m) diameter concrete-lined penstocks (tunnels). Just before reaching the power house and its eight turbines, each penstock splits into four branches.

The power house is about 2,500 ft (760 m) downstream of the dam and utilizes two surge tanks for sudden rises in water pressure from the two penstocks. Each surge tank has a 80 ft (24 m) diameter expansion chamber and is about 40 ft (12 m) higher than the actual dam structure. The surge tanks protect the penstocks and turbines from water hammer which would occur if the turbine gates were quickly closed and water pressure suddenly increases.

Cultural and political significance

The construction of the Daniel-Johnson Dam and the Manic-Outardes complex happened in a larger social and political context of the Quiet Revolution in Quebec, a time when recently nationalized "Hydro-Québec rapidly becomes a symbol of the new Quebec nationalism and of the new economic strategy of the State", explains historian Paul-André Linteau.

This newfound euphoric attitude is found in newspapers of the period. In a series of papers published in Montreal's La Presse, Renaude Lapointe calls Hydro-Québec a "colossus on the march". In this context, the construction of the complex is closely followed by the public and becomes part of the popular culture of that era.

For instance, chansonnier Georges Dor penned his huge 1966 hit, La Manic. The song tells the story of a construction worker at the remote jobsite who describes his loneliness to his wife in words that captured the collective imagination of the Quebec public.

Belgian-born novelist Henri Vernes was also inspired by the gigantic project and made the Manic-5 project the setting of one of his Bob Morane adventure novels. Terreur à la Manicouagan is about an attempt by arch-villain Roman Orgonetz to breach the dam by destroying the surge chamber by remote-controlled detonation, a plan foiled by Morane and his Scottish sidekick, Bill Ballantine. Prior to writing his novel, Vernes spent some time at the worksite, at the invitation of the Quebec government and Hydro-Québec. The book was launched in 1965 at the Hydro-Québec Building in Montreal.

Other notable visitors to the construction site included famous cartoonist Hergé, who left an original drawing of his characters Tintin the reporter and his dog Snowy posing in front of a ligne claire depiction of the dam. A colorized version of the autographed drawing is featured at the Jean-Lesage generating station visitor's center, 23 kilometres (14 mi) north of Baie-Comeau.

In the late 60s, a Montreal-based public relations officer with Renault Canada, Jacques About, was asked by the company to study the feasibility of introducing the Renault Alpine sports car into Canada. About's survey showed some potential but the manufacturer ultimately decided against introducing the model in its North American offerings.

The Montrealer decided to leave the company and start his own manufacturing business, Automobiles Manic Inc., in 1968. After gaining financial support from public and private backers, including Bombardier and Steinberg's supermarkets, his company started building a 2-seat sport coupe based on the Alpine A110, the Manic GT. Due to disappointing sales, the company stopped making the car in 1971.

The hydroelectric project also gave its name to Montreal's first foray into professional soccer, the short-lived Montreal Manic. The franchise competed for three seasons in the North American Soccer League in the early 80s. In March 2000, the Canada Post Corporation issued a 46 cent stamp featuring the Dam as part of a 4-stamp Millenium Collection sheet depicting "engineering and technological marvels".

10

CANAL IRRIGATION IN INDIA

Canal irrigation is the most important form of irrigation in India. It is cheaper. About half of 165.97 lakh hectare land was irrigated by canals during 2008-09 from which half of irrigated land concentrated in the Northern plain. If we calculate it statewise, then 91.72 % of irrigated area in Jammu Kashmir, 66.24 % in Chhattisgarh, 64.7 % in Odisha, 44.28% in Haryana and 34.63 % in Andhra Pradesh is irrigated by canals.

The maximum part of the total irrigated area of the country by canals is in Uttar Pradesh. Other major states where irrigation is done by canals are Madhya Pradesh, Andhra Pradesh, Haryana, Punjab and Bihar. There are two types- Inundation Canal and Perennial Canals.

SOURCES AND METHODS OF IRRIGATION

The monsoonal rainfall in India is concentrated only in four months and more than 50% of the net sown area is rainfed only. Irrigation is thus essential to overcome spatial and temporal variation of rainfall.

Archaeological and historical records show that from ancient times we have been constructing sophisticated hydraulic structures like dams built of stone rubble, reservoirs or lakes, embankments and canals for irrigation. Not surprisingly, we have continued this tradition in modern India by building dams in most of our river basins. Before we look at these methods of irrigation in detail, let's have a look at some of the hydraulic structures used in ancient India!

Some Hydraulic Structures used in Ancient India:

In the first century BC, Sringaverapura near Allahabad had sophisticated water harvesting system channelling the flood water of the river Ganga.

During the time of Chandragupta Maurya, dams, lakes and irrigation systems were extensively built.

Evidences of sophisticated irrigation works have also been found in Kalinga (Orissa), Nagarjunakonda (Andhra Pradesh), Bennur (Karnataka), Kolhapur (Maharashtra), etc.

In the eleventh century, Bhopal Lake, one of the largest artificial lakes of its time was built.

In the 14th century, the tank in Hauz Khas, Delhi was constructed by Iltutmish for supplying water to the Siri Fort Area.

Coming back to irrigation in the present day India, let's look at some important facts and figures before we move forward:

Some important facts and figures:

The net irrigated area = 66.1 million hectares.

Total/Gross Irrigated Area = 92.6 million hectares.

Irrigation Intensity in India = (Gross Irrigated Area ÷Gross Sown Area) * 100

= (92.6 ÷ 194.4) *100

= 47.6%

More than 50% of the country's cropped area depends exclusively on rainfall, most of which is concentrated in a few months of the year. Even where the annual overall precipitation is high, the available moisture is not adequate to support multiple cropping.

Ultimate Irrigation Potential

As seen in the above figures, only about 66mha i.e. 47.6% of the net sown area is estimated to be irrigated. There is a need to bring more cropped area under assured irrigation so as to increase agricultural productivity and production.

The total ultimate irrigation potential of the country has been estimated as 140mha, with about 76 mha from surface water sources and about 64mha from groundwater sources.

Irrigation – Sources and Methods

The main sources of irrigation in India are:

Canals

Wells (and tubewells)

Tanks

The relative importance of these has been changing from time to time. Let's look at these in detail:

1. Canal Irrigation:

A canal is an artificial watercourse constructed for water supply and irrigation.

There are two types of canals:

Inundation Canals – These are taken out from the rivers without any regulating system like weirs etc at their head. Such canals are useful only during the rainy season

Perennial Canals – These are those which are taken off from perennial rivers by constructing a barrage across the river. Most of the canals at present in India are perennial.

Canals can be an effective source of irrigation in areas of low relief, deep fertile soils, perennial source of water and an extensive command area. Therefore the main concentration of canal irrigation is in the northern plains.

The canals are practically absent from the peninsular plateau region because of rocky terrain. However, the coastal and the delta regions in South India have some canals for irrigation.

The percentage of canal irrigation area to total irrigated area in the country has fallen from about 40% in 1950-51 to less than 25% at present.

The states UP, Punjab, Haryana, Rajasthan and Bihar account for about 60% of the canal irrigated area in the country.

Merits of canal irrigation:

Perennial Source

Provides safety from droughts

Brings fertile sediments to the fields

Economical to serve a large area

Demerits:

Canal water soaks into the ground and leads to water logging, increases salinization, and leads to marshy conditions leading to malaria and flooding

Wastage of water.

2. Wells (and Tube Wells)

A well is a hole dug in the ground to obtain the subsoil water. An ordinary well is about 3-5 metres deep but deeper wells up to 15 metres are also dug.

This method of irrigation has been used in India from time immemorial. Various methods are used to lift the ground water from the well. Some of the widely used methods are the persian wheel, reht, charas or mot, and dhinghly (lever) etc.

A tube well is a deeper well (generally over 15 metres deep) from which water is lifted with the help of a pumping set operated by an electric motor or a diesel engine.

Well irrigation is gradually giving way to energized tube wells. But there are many wells still in use where electricity is not available or the farmers are too poor t0 afford diesel oil.

This method of irrigation is popular in those areas where sufficient sweet ground water is available.

It is particularly suitable in areas with permeable rock structure which allows accumulation of ground water through percolation. Therefore wells are seen more in areas with alluvial soil, regur soil, etc. and less seen in rocky terrain or mountainous regions.

These areas include a large part of the great northern plains, the deltaic regions of the Mahanadi, the Godavari, the Krishna and the Cauvery, parts of the Narmada and the Tapi valleys and the weathered layers of the Deccan trap and crystalline rocks and the sedimentary zones of the peninsula

However, the greater part of peninsular India is not suitable for well irrigation due to rocky structure, uneven surface and lack of underground water.

Large dry tracts of Rajasthan, the adjoining parts of Punjab, Haryana and Gujarat and some parts of Up have brackish ground water which is not fit for irrigation and human consumption and hence unsuitable for well irrigation

At present irrigation from wells and tubewells accounts for more than 60% of the net irrigated area in the country.

UP has the largest area under well irrigation which accounts for 28% of the well irrigated area of the country. U.P., Rajasthan, Punjab, Madhya Pradesh, Gujarat, Bihar and Andhra Pradesh account for about three-fourths of the total well-irrigated area

Merits of well irrigation

Simplest

Cheapest

Well is an independent source of irrigation and can be used as and when the necessity arises. Canal irrigation, on the other hand, is controlled by other agencies and cannot be used at will.

Some ground water salts are useful for crops

Does not lead to salinization and flooding problems

There is a limit to the extent of canal irrigation beyond the tail end of the canal while a well can be dug at any convenient place.

Demerits

Only limited area can be irrigated. Normally, a well can irrigate 1 to 8 hectares of land.

Not suitable for dry regions

Overuse may lead to lowering of water table

3. Tank irrigation

A tank is a reservoir for irrigation, a small lake or pool made by damming the valley of a stream to retain the monsoon rain for later use.

It accounts for approximately 3% of the net irrigated area in India.

Tank Irrigation is popular in the peninsular plateau area where Andhra Pradesh and Tamil Nadu are the leading states.

Andhra Pradesh has the largest area (29%) of tank irrigation in India followed by Tamil nadu (23%).

It is practised mainly in the peninsular region due to the following reasons:

The undulating relief and hard rocks make it difficult to dig canals and wells

There is little percolation of water due to hard rock structure and ground water is not available in large quantities.

Most of the rivers are seasonal; there are many streams which become torrential during the rainy season – so the only way to use this water is to impound it by constructing bunds and building tanks. Also, it is easy to collect rainwater in natural or artificial pits because of impermeable rocks.

Scattered nature of agricultural fields

Merits

Most of the tanks are natural and do not involve cost for their construction

Independent source for an individual farmer or a small group of farmers

longer life span

can be used for fishing also

Demerits

Depends on rain and these tanks may dry up during the dry season

Silting of their beds

Require large areas

Evaporation losses

Sometimes there might be a need to lift the water to take it to the field

TYPES OF ARTIFICIAL WATERWAYS

Canals are created in one of three ways, or a combination of the three, depending on available water and available path:

- A canal can be created where no stream presently exists. The body of the canal is either dug or the sides are created by piling dirt, stone, concrete, or other building materials. The water for the canal must be provided from an external source like other streams or reservoirs. Examples include canals that connect valleys over a higher body of land, like Canal du Midi and Canal de Briare.
- A stream can be canalized to make its navigable path more predictable and easier to maneuver. Canalization modifies the stream to more safely carry traffic by controlling the flow of the stream with dredging, damming, and modifying its path. Examples include Basse Saône, Canal de Mines de Fer de la Moselle, and Aisne River. Riparian zone restoration may be required.
- When a stream is too difficult to modify with canalization, a second stream can be created next to the existing stream. This is called a lateral canal. The existing stream usually acts as the water source and its banks provide a path for the new body. Examples include the Chesapeake and Ohio Canal, Canal latéral à la Loire, Garonne Lateral Canal, and Canal latéral à l'Aisne.

Smaller transportation canals can carry barges or narrowboats, while ship canals allow seagoing ships to travel to an inland port (e.g.: Manchester Ship Canal), or from one sea or ocean to another (e.g.: Caledonian Canal, Panama Canal).

Features

At their simplest, canals consist of a trench filled with water. Depending on the stratum the canal passes through, it may be necessary to line the cut with some form of watertight material such as clay or concrete. When this is done with clay this is known as puddling.

Canals need to be level, and while small irregularities in the lie of the land can be dealt with through cuttings and embankments, for larger deviations, other approaches have been adopted. The most common is the pound lock which consists of a chamber within which the water level can be raised or lowered connecting either two pieces of canal at a different level or the canal with a river or the sea. When there is a hill to be climbed, flights of many locks in short succession may be used.

Prior to the development of the pound lock in 984AD in China by Chhaio Wei-Yo and later in Europe in the 15th century, either flash locks consisting of a single gate were used or ramps, sometimes equipped with rollers, were used to change level. Flash locks were only practical where there was plenty of water available.

Locks use a lot of water, so builders have adopted other approaches. These include boat lifts, such as the Falkirk wheel, which use a caisson of water in which boats float while being moved between two levels; and inclined planes where a caisson is hauled up a steep railway.

To cross a stream or road, the solution is usually to bridge with an aqueduct. To cross a wide valley (where the journey delay caused by a flight of locks at either side would be unacceptable) the centre of the valley can be spanned by an aqueduct - a famous example in Wales is the Pontcysyllte aqueduct across the valley of the River Dee.

Another option for dealing with hills is to tunnel through them. An example of this approach is the Harecastle Tunnel on the Trent and Mersey Canal. Tunnels are only practical for smaller canals.

Some canals attempted to keep changes in level down to a minimum. These canals known as contour canals would take longer winding routes, along which the land was a uniform altitude. Other generally later canals took more direct routes requiring the use of various methods to deal with the change in level.

Canals have various features to tackle the problem of water supply. In some cases such as the Suez Canal the canal is simply open to the sea. Where the canal is not at sea level a number of approaches have been adopted. Taking water from existing rivers or springs was an option in some cases, sometimes supplemented by other methods to deal with seasonal variations in flow. Where such sources were unavailable, reservoirs, either separate from the canal, or built into its course, and back pumping was used to provide the required water. In other cases water pumped from mines was used to feed the canal. In certain cases extensive "feeder canals" were built to bring water from sources located far from the canal.

Where large amounts of goods are loaded or unloaded such as at the end of a canal a canal basin may be built. This would normally be a section of water wider than the general canal. In some cases the canal basins contain wharfs and cranes to assist with movement of goods.

When a section of the canal needs to be sealed off so it can be drained for maintenance stop planks are frequently used. These consist of planks of wood placed across the canal to form a dam. They are generally placed in pre existing grooves in the canal bank. On more modern canals, "guard locks" or gates were sometimes placed to allow a section of canal to be quickly closed off, either for maintenance, or to prevent a major loss of water due to a canal breach.

The oldest known canals were irrigation canals, built in Mesopotamia circa 4000 BC, in what is now modern day Iraq and Syria. The Indus Valley Civilization, in Pakistan and North India, (circa 2600 BC) had sophisticated irrigation and storage systems developed, including the reservoirs built at Girnar in 3000 BC. In Egypt, canals date back at least to the time of Pepi I Meryre (reigned 2332-2283 BC), who ordered a canal built to bypass the cataract on the Nile near Aswan.

In ancient China, large canals for river transport were established as far back as the Warring States (481-221 BC), the longest one of that period being the Hong Gou (Canal of the Wild Geese), which according to the ancient historian Sima Qian connected the old states of Song, Zhang, Chen, Cai, Cao, and Wei.

By far the longest canal was the Grand Canal of China, still the longest canal in the world today. It is 1,794 kilometres (1,115 mi) long and was built to carry the Emperor Yang Guang between Beijing and Hangzhou. The project began in 605 and was completed in 609, although much of the work combined older canals, the oldest section of the canal existing since at least 486 BC. Even in its narrowest urban sections it is rarely less than 30 metres (98 ft) wide. Greek engineers were the first to use canal locks, by which they regulated the water flow in the Ancient Suez Canal as early as the 3rd century BC.

Canals in the Middle Ages

In the Middle Ages, water transport was cheaper and faster than transport overland. This was because roads were unpaved and in poor condition and greater amounts could be transported by ship. The first artificial canal in Christian Europe was the Fossa Carolina built at the end of the 8th Century under personal supervision of Charlemagne. More lasting and of more economic impact were canals like the Naviglio Grande built between 1127 and 1257, the most important of the lombard "navigli", Later, canals were built in the Netherlands and Flanders to drain the polders and assist the transportation of goods.

Canal building was revived in this age because of commercial expansion from the 12th century AD. River navigations were improved progressively by the use of single, or flash locks. Taking boats through these used large amounts of water leading to conflicts with watermill owners and to correct this, the pound or chamber lock first appeared, in 10th century AD in China and in Europe in 1373 in Vreeswijk, Netherlands. Another important development was the mitre gate which was probably introduced in Italy by Bertola da Novate in the sixteenth century. This allowed wider gates and also removed the height restriction of guillotine locks.

To break out of the limitations caused by river valleys, the first summit level canals were developed with the Grand Canal of China in 581-617 AD whilst in Europe the first, also using single locks, was the Stecknitz Canal in Germany in 1398. The first to use pound locks was the Briare Canal connecting the Loire and Seine (1642), followed by the more ambitious Canal du Midi (1683) connecting the Atlantic to the Mediterranean. This included a staircase of 8 locks at Béziers, a 157 metres (515 ft) tunnel and three major aqueducts.

Canal building progressed steadily in Germany in the 17th and 18th centuries with three great rivers, the Elbe, Oder and Weser being linked by canals. In post-Roman Britain, the first canal built appears to have been the Exeter Canal, which opened in 1563. The oldest canal built for industrial purposes in North America is Mother Brook in Dedham, MA. It was constructed in 1639 to provide water power for mills. In Russia, the Volga-Baltic Waterway, a nationwide canal system connecting the Baltic and Caspian seas via the Neva and Volga rivers, was opened in 1718.

Industrial revolution

Canals were important for industrial development. The greatest stimulus to canal system building came from the Industrial Revolution with its need for cheap transport of unprecedented quantities of raw materials and manufactured items.

In Europe, particularly Britain and Ireland, and then in the young United States and the Canadian colonies, inland canals preceded the development of railroads during the earliest phase of the Industrial Revolution. The opening of the Sankey Canal in 1757, followed by the Bridgewater Canal in 1761, which halved the price of coal in Liverpool and Manchester, respectively, triggered a period of "canal mania" in Britain so that between 1760 and 1820 over one hundred canals were built.

The Blackstone Canal in Massachusetts and Rhode Island fulfilled a similar role in the early industrial revolution between 1828-48. The Blackstone Valley was a major contributor of the American Industrial Revolution where Samuel Slater built his first mill.

In addition to their transportation purposes, parts of the United States, particularly in the Northeast, had enough fast-flowing rivers that water power was the primary means of powering factories (usually textile mills) until after the American Civil War. For example, Lowell, Massachusetts, considered to be "The Cradle of the American Industrial Revolution," has 6 miles (9.7 km) of canals, built from around 1790 to 1850, that provided water power and a means of transportation for the city. The output of the system is estimated at 10,000 horsepower. Other cities with extensive power canal systems include Lawrence, Massachusetts, Holyoke, Massachusetts, Manchester, New Hampshire, and Augusta, Georgia.

The 19th century

Competition from the railway network from the 1830s, and in the 20th century the roads, made the smaller canals obsolete for most commercial transportation, and many of the British canals fell into decay. Only the Manchester Ship Canal and the Aire and Calder Canal bucked this trend. Yet in other countries canals grew in size as construction techniques improved. During the 19th century in the US, the length of canals grew from 100 miles (161 km) to over 4,000, with a complex network making the Great Lakes navigable, in conjunction with Canada, although some canals were later drained and used as railroad rights-of-way.

In the United States, navigable canals reached into isolated areas and brought them in touch with the world beyond. By 1825 the Erie Canal, 363 miles (584 km) long with 82 locks, opened up a connection from the populated Northeast to the Great Lakes. Settlers flooded into regions serviced by such canals, since access to markets was available. The Erie Canal (as well as other canals) was instrumental in lowering the differences in commodity prices between these various markets across America.

The canals caused price convergence between different regions because of their reduction in transportation costs, which allowed Americans to ship and buy goods from farther distances for much lower prices compared to before. Ohio built many miles of canal, Indiana had working canals for a few decades, and the Illinois and Michigan Canal connected the Great Lakes to the Mississippi River system until replaced by a channelized river waterway.

Three major canals with very different purposes were built in what is now Canada. The first Welland Canal, which opened in 1829 between Lake Ontario and Lake Erie, bypassing Niagara Falls and the Lachine Canal (1825) which allowed ships to skirt the nearly impassable rapids on the St. Lawrence River at Montreal were built for commerce.

The Rideau Canal, completed in 1832, connects Ottawa, on the Ottawa River to Kingston, Ontario on Lake Ontario. The Rideau Canal was built as a result of the War of 1812 to provide military transportation between the British colonies of Upper Canada and Lower Canada as an alternative to part of the St. Lawrence River which was susceptible to blockade by the United States.

In France, a steady linking of all the river systems-Rhine, Rhône, Saône and Seine-and the North Sea was boosted in 1879 by the establishment of the Freycinet gauge which specified the minimum size of locks so that canal traffic doubled in the first decades of the 20th century.

Many notable sea canals were completed in this period, starting with the Suez Canal (1869), and the Kiel Canal (1897), which carries tonnage many times that of most other canals, though the Panama Canal was not opened until 1914.

In the 19th century, a number of canals were built in Japan including the Biwako canal and the Tone canal. These canals were partially built with the help of engineers from the Netherlands and other countries.

Modern uses

Large scale ship canals such as the Panama Canal and Suez Canal continue to operate for cargo transportation; as do European barge canals. Due to globalization, they are becoming increasingly important, resulting in expansion projects such as the Panama Canal expansion project.

The narrow early industrial canals, however, have ceased to carry significant amounts of trade and many have been abandoned to navigation, but may still be used as a system for transportation of untreated water. In some cases railways have been built along the canal route, an example being the Croydon Canal.

A movement that began in Britain and France to use the early industrial canals for pleasure boats, such as hotel barges, has spurred rehabilitation of stretches of historic canals. In some cases abandoned canals such as the Kennet and Avon Canal have been restored and are now used by pleasure boaters. In Britain canalside housing has also proven popular in recent years.

Canals have found another use in the 21st century, as easements for the installation of fibre optic telecommunications network cabling avoiding having them buried in roadways while facilitating access and reducing the hazard of being damaged from digging equipment.

Canals are still used to provide water for agriculture. An extensive canal system exists within the Imperial Valley in the Southern California desert to provide irrigation to agriculture within the area.

Canals are so deeply identified with Venice that many canal cities have been nicknamed "the Venice of…". The city is built on marshy islands, with wooden piles supporting the buildings, so that the land is man-made rather than the waterways. The islands have a long history of settlement; by the 12th century, Venice was a powerful city state. Amsterdam was built in a similar way, with buildings on wooden piles. It became a city around 1300.Other cities with extensive canal networks include: Delft, Haarlem and Leiden in the Netherlands, Brugge in Flanders, Belgium, Birmingham in England - which has 35 miles of canals, to Venice's 26 miles -, Saint Petersburg in Russia, Hamburg in Germany, Berlin in Germany , Fort Lauderdale, Florida, and Cape Coral, Florida in the United States.

Liverpool Maritime Mercantile City is a UNESCO World Heritage Site near the centre of Liverpool, England, where a system of intertwining waterways and docks now being developed for mainly residential and leisure use.

Canal Estates (commonly known as bayous) are a form of subdivision popular in cities like Miami, Florida, Texas City, Texas and the Gold Coast, Queensland; the Gold Coast has over 700 km of residential canals. Wetlands are difficult areas upon which to build housing estates, so dredging part of the wetland down to a navigable channel provides fill to build up another part of the wetland above the flood level for houses. Land is built up in a finger pattern that provides a suburban street layout of waterfront housing blocks.

Inland canals have often had boats specifically built for them. An example of this is the British narrowboat, which is up to 72 feet (21.95 m) long and 7 feet (2.13 m) wide and was primarily built for British Midland canals. In this case the limiting factor was the size of the locks. This is also the limiting factor on the Panama canal where Panamax ships are limited to a length of 294.1 m (965 ft) and a width of 32.3 m (106 ft).

For the lockless Suez Canal the limiting factor for Suezmax ships is generally draft, which is limited to 16 m (52.5 ft). At the other end of the scale, tub-boat canals such as the Bude Canal were limited to boats of under 10 tons for much of their length due to the capacity of their inclined planes or boat lifts. Most canals have a limit on height imposed either by bridges or tunnels.

RIDGE

A ridge is a geological feature that features a chain of mountains or hills that are of a continuous elevated crest for some distance. Ridges are usually termed hills or mountains as well, depending on size. There are several main types of ridges:

- Dendritic ridge: In typical dissected plateau terrain, the stream drainage valleys will leave intervening ridges. These are by far the most common ridges. These ridges usually represent slightly more erosion resistant rock, but not always - they often remain because there were more joints where the valleys formed, or other chance occurrences. This type of ridge is generally somewhat random in orientation, often changing direction frequently, often with knobs at intervals on the ridge top.

- Stratigraphic ridge: In places such as the Ridge-and-valley Appalachians, long, even, straight ridges are formed because they are the uneroded remaining edges of the more resistant strata that were folded laterally. Similar ridges have formed in places such as the Black Hills, where the ridges form concentric circles around the igneous core. Sometimes these ridges are called "hogback ridges".
- Oceanic spreading ridge: In tectonic spreading zones around the world, such as at the Mid-Atlantic Ridge, the volcanic activity forming new plate boundary forms volcanic ridges at the spreading zone. Isostatic settling and erosion gradually reduce the elevations moving away from the zone.
- Crater ridges: Large meteorite strikes typically form large impact craters bordered by circular ridges.
- Volcanic caldera ridges: Large volcanoes often leave collapsed central calderas that are bordered by circular ridges.
- Fault ridges: Faults often form escarpments. Sometimes the tops of the escarpments form not plateaus, but slope back so that the edges of the escarpments form ridges.
- Dune ridges: In areas of large-scale dune activity, certain types of dunes result in sand ridges.
- Moraines and eskers: Glacial activity may leave ridges in the form of moraines and eskers. An arête is a thin ridge of rock that is formed by glacial erosion.
- Volcanic subglacial ridges: Many subglacial volcanoes create ridgelike formations when lava erupts through a thick glacier or ice sheet.

Tectonic Uplift

Tectonic uplift is a geological process most often caused by plate tectonics which increases elevation. The opposite of uplift is subsidence, which results in a decrease in elevation. Uplift may be orogenic or isostatic. Orogenic uplift is the result of tectonic-plate collisions and results in mountain ranges or a more modest uplift over a large region. The Himalayas were (and are still being) formed by the collision of two continental plates, the Indian and Eurasian Plates. This ongoing collision produced the Tibetan Plateau as well as the Himalayas and associated ranges.

Ozark Plateau is a broad uplifted area which resulted from the Ouachita Orogeny to the south during the Permian Period. Another related uplift is the

Llano Uplift in Texas, a geographical location named after its uplift features. The Colorado Plateau with its spectacular scenic canyons, the Grand Canyon, is also the result of broad tectonic uplift followed by river erosion.

Isostatic uplift

Isostatic uplift includes the gradual uplift following rapid erosional removal of material from a mountain range. The land rises as a result of the removal of the weight. Another example of isostatic uplift is post-glacial rebound following the melting of continental glaciers and ice sheets. The Hudson Bay region of Canada and the Great Lakes of Canada and the United States are currently undergoing gradual rebound as a result of the melting of the ice sheets 10,000 years ago.

Coral islands

In a few cases, tectonic uplift can be seen in the cases of coral islands. This is evidenced by the presence of various oceanic islands composed entirely of coral, which otherwise appear to be high islands (i.e., islands of volcanic origin). Examples of such islands are found in the Pacific, notably the three phosphate islets, Nauru, Makatea, and Banaba as well as Maré and Lifou in New-Caledonia, Fatu Huku in the Marquesas Islands and Henderson Island in the Pitcairn Islands. The uplift of these islands is the result of the movement of oceanic tectonic plates. Sunken islands or guyots with their coral reefs are the result of crustal subsidence as the oceanic plate carries the islands to deeper or lower oceanic crust areas.

Waterlogging

It refers to the saturation of soil with water. Soil may be regarded as waterlogged when the water table of the groundwater is too high to conveniently permit an anticipated activity, like agriculture.

In agriculture, various crops need air (specifically, oxygen) to a greater or lesser depth in the soil. Waterlogging of the soil stops air getting in. How near the water table must be to the surface for the ground to be classed as waterlogged, varies with the purpose in view. A crop's demand for freedom from waterlogging may vary between seasons of the year, as with the growing of rice (Oryza sativa).

In irrigated agricultural land, waterlogging is often accompanied by soil salinity as waterlogged soils prevent leaching of the salts imported by the irrigation water.

From a gardening point of view, waterlogging is the process whereby the soil blocks off all water and is so hard it stops air getting in and it stops oxygen from getting in.

Soil salinity control

Soil salinity control relates to controlling the problem of soil salinity and reclaiming salinized agricultural land. The aim of soil salinity control is to prevent soil degradation by salinization and reclaim already salty (saline) soils. Soil reclamation is also called soil improvement, rehabilitation, remediation, recuperation, or amelioration.

The primary man-made cause of salinization is irrigation. River water or groundwater used in irrigation contains salts, which remain behind in the soil after the water has evaporated.

The primary method of controlling soil salinity is to permit 10-20% of the irrigation water to leach the soil, be drained and discharged through an appropriate drainage system. The salt concentration of the drainage water is normally 5 to 10 times higher than that of the irrigation water, thus salt export matches salt import and it will not accumulate.

Salty (saline) soils are soils that have a high salt content. The predominant salt is normally sodium chloride (NaCl, "table salt"). Saline soils are therefore also sodic soils but there may be sodic soils that are not saline, but alkaline.

Salty soils are a common feature and an environmental problem in irrigated lands in arid and semi-arid regions. They have poor or little crop production. The problems are often associated with high water tables, caused by a lack of natural subsurface drainage to the underground. Poor subsurface drainage may be caused by insufficient transport capacity of the aquifer or because water cannot exit the aquifer for instance, if it is situated in a topographical depression.

Worldwide, the major factor in the development of saline soils is a lack of precipitation. Most naturally saline soils are found in (semi)arid regions and climates of the globe.

Primary cause

The primary cause of man-made salinization is the salt brought in with irrigation water. All irrigation water derived from rivers or groundwater, however 'sweet', contains salts that remain behind in the soil after the water has evaporated, which may create the environmental issue.

For example, assuming irrigation water with a low salt concentration of 0.3 g/l (equal to 0.3 kg/m^3 corresponding to an electric conductivity of about 0.5 dS/m) and a modest annual supply of irrigation water of 10,000 m^3/ha (almost 3 mm/day) brings 3,000 kg salt/ha each year. In the absence of sufficient natural drainage (as in waterlogged soils) and without a proper leaching and drainage program to remove salts, this would lead to a high soil salinity and reduced crop yields in the long run.

Much of the water used in irrigation has a higher salt content than in this example, which is compounded by that fact that many irrigation projects use a far greater annual supply of water. Sugar cane, for example, needs about 20000 m3/ha of water per year. As a result, irrigated areas often receive more than 3,000 kg/ha of salt per year and some receive as much as 10,000 kg/ha/year.

Secondary cause

The secondary cause of salinization is waterlogging in irrigated land.

Irrigation causes changes to the natural water balance of irrigated lands. Large quantities of water in irrigation projects are not consumed by plants and must go somewhere. In irrigation projects it is impossible to achieve 100% irrigation efficiency where all the irrigation water is consumed by the plants. The maximum attainable irrigation efficiency is about 70% but usually it is less than 60%. This means that minimum 30%, but usually more than 40% of the irrigation water is not evaporated and it must go somewhere.

Most of the water lost this way is stored underground which can change the original hydrology of local aquifers considerably. Many aquifers cannot absorb and transport these quantities of water and, the water table rises leading to water logging.

Waterlogging causes three problems:

- The shallow water table and lack of oxygenation of the root zone reduces the yield of most crops
- It leads to an accumulation of salts brought in with the irrigation water as their removal through the aquifer is blocked
- With the upward seepage of groundwater more salts are brought into the soil and the salination is aggravated

Aquifer conditions in irrigated land and the groundwater flow have an important role in soil salinization, as illustrated here :

- Illustration of the influence of aquifer conditions on soil salinization in irrigated land
- Soil salinization in the lower parts of undulating land with a good aquifer

- Soil salinization in the unirrigated parts of flat land with a good aquifer
- Soil salinization in irrigated flat land without an aquifer

Salt affected area

Normally, the salinization of agricultural land affects a considerable area of irrigation projects, on the order of 20 to 30%. When the agriculture in such a fraction of the land is abandoned, a new salt and water balance is attained, a new equilibrium is reached, and the situation becomes stable.

In India alone, thousands of square kilometres have been severely salinized. China and Pakistan do not lag much behind (perhaps China has even more salt affected land than India). A regional distribution of the 3,230,000 km^2 of saline land world wide is shown in the following table derived from the FAO/UNESCO Soil Map of the World.

Spatial variation

Although the principles of the processes of salinization are fairly easy to understand, it is more difficult to explain why certain parts of the land suffer from the problems and other parts do not, or to predict accurately which part of the land will fall victim. The main reason for this is the variation of natural conditions in time and space, the usually uneven distribution of the irrigation water, and the seasonal or yearly changes of agricultural practices.

Only in lands with undulating topography the explanation and prediction is pretty simple: the depressional areas will degrade the most.

The preparation of salt and water balances for distinguishable sub-areas in the irrigation project, or the use of agro-hydro-salinity models, can be helpful in explaining or predicting the extent and severity of the problems.

Diagnosis

Measurement

Soil salinity is measured as the salt concentration of the soil solution in tems of g/l or electric conductivity (EC) in dS/m. The relation between these two units is about 5/3 : y g/l => 5y/3 dS/m. Sea water may have a salt concentration of 30 g/l (3%) and an EC of 50 dS/m.

The standard for the determination of soil salinity is from an extract of a saturated paste of the soil, and the EC is then written as ECe. The extract is obtained by centrifugation. The salinity can more easily be measured, without centrifugation, in a 2:1 or 5:1 water:soil mixture (in terms of g water per g dry soil) than from a saturated paste. The relation between ECe and EC2:1 is about 4, hence : ECe = 4 EC1:2.

Classification

Soils are considered saline when the ECe > 4. When 4 < ECe < 8, the soil is called slightly saline, when 8 < ECe < 16 it is called (moderately) saline, and when ECe > 16 severely saline.

Crop tolerance

Sensitive crops loose their vigor already in slightly saline soils, most crops are negatively affected by (moderately) saline soils, and only salinity resistant crops thrive in severely saline soils. The University of Wyoming and the Government of Alberta report data on the salt tolerance of plants.

PRINCIPLES OF SALINITY CONTROL

Drainage is the primary method of controlling soil salinity. The system should permit a small fraction of the irrigation water (about 10 to 20 percent, the drainage or leaching fraction) to be drained and discharged out of the irrigation project.

In irrigated areas where salinity is stable, the salt concentration of the drainage water is normally 5 to 10 times higher than that of the irrigation water. Salt export matches salt import and salt will not accumulate.

When reclaiming already salinized soils, the salt concentration of the drainage water will initially be much higher than that of the irrigation water (for example 50 times higher). Salt export will greatly exceed salt import, so that with the same drainage fraction a rapid desalinization occurs. After one or two years, the soil salinity is decreased so much, that the salinity of the drainage water has come down to a normal value and a new, favorable, equilibrium is reached.

In regions with pronounced dry and wet seasons, the drainage system may be operated in the wet season only, and closed during the dry season. This practice of checked or controlled drainage saves irrigation water.

The discharge of salty drainage water may pose environmental problems to downstream areas. The environmental hazards must be considered very carefully and, if necessary mitigating measures must be taken. If possible, the drainage must be limited to wet seasons only, when the salty effluent inflicts the least harm.

Drainage systems

The drainage system designed to evacuate salty water also lowers the water table. To reduce the cost of the system, the lowering must be reduced to a minimum. The highest permissible level of the water table (or the shallowest permissible depth) depends on the irrigation and agricultural practices and kind of crops.

In many cases a seasonal average water table depth of 0.6 to 0.8 m is deep enough. This means that the water table may occasionally be less than 0.6 m (say 0.2 m just after an irrigation or a rain storm). This automatically implies that, in other occasions, the water table will be deeper than 0.8 m (say 1.2 m). The fluctuation of the water table helps in the breathing function of the soil while the expulsion of carbon dioxide (CO2) produced by the plant roots and the inhalation of fresh oxygen (O2) is promoted.

The establishing of a not too deep water table offers the additional advantage that excessive field irrigation is discouraged, as the crop yield would be negatively affected by the resulting elevated water table, and irrigation water may be saved.

The statements made above on the optimum depth of the watertable are very general, because in some instances the required water table may be still shallower than indicated (for example in rice paddies), while in other instances it must be considerably deeper (for example in some orchards). The establishment of the optimum depth of the water table is in the realm of agricultural drainage criteria.

Soil leaching

The unsaturated zone or vadose zone of the soil below the soil surface and the watertable is subject to four main hydrological inflow and outflow factors :

- Infiltration of rain and irrigation water (Irr) into the soil through the soil surface (Inf) :
- Inf = Rain + Irr
- Evaporation of soil water through plants and directly into the air through the soil surface (Evap)
- Percolation of water from the unsaturated zone soil into the groundwater through the watertable (Perc)
- Capillary rise of groundwater moving by capillary suction forces into the unsaturated zone(Cap)

In steady state (i.e. the amount of water stored in the unsaturated zone does not change in the long run) the water balance of the unsaturated zone reads: Inflow = Outflow, thus:

- Inf + Cap = Evap + Perc or :
- Irr + Rain + Cap = Evap + Perc and the salt balance is
- Irr.Ci + Cap.Cc = Evap.Fc.Ce + Perc.Cp + Ss

where Ci is the salt concentration of the irrigation water, Cc is the salt concentration of the capillary rise, equal to the salt concentration of the upper part of the groundwater body, Fc is the fraction of the total evaporation transpired by plants, Ce is the salt concentration of the water taken up by the plant roots, Cp is the salt concentration of the percolation water, and Ss is the increase of salt storage in the unsaturated soil. This assumes that the rainfall contains no salts. Only along the coast this may not be true. Further it is assumed that no runoff or surface drainage occurs.

The amount of salts removed by plants (Evap.Fc.Ce) is usually negligibly small: Evap.Fc.Ce = 0

The salt concentration Cp can be taken as a part of the salt concentration of the soil in the unsaturated zone (Cu) giving: Cp=Le.Cu, where Le is the leaching efficiency. The leaching efficiency is often in the order of 0.7 to 0.8, but in poorly structured, heavy clay soils it may be less. In the Leziria Grande polder in the delta of the Tagus river in Portugal it was found that the leaching efficiency was only 0.15.

Assuming that one wishes to avoid the soil salinity to increase and maintain the soil salinity Cu at a desired level Cd we have:

Ss = 0, Cu = Cd and Cp = Le.Cd. Hence the salt balance can be simplified to:

- Perc.Le.Cd = Irr.Ci + Cap.Cc

Setting the amount percolation water required to fulfill this salt balance equal to Lr (the leaching requirement) it is found that:

- Lr = (Irr.Ci + Cap.Cc) / Le.Cd .

Substituting herein Irr = Evap + Perc ? Rain ? Cap and re-arranging gives :

- Lr = [(Evap?Rain).Ci + Cap(Cc?Ci)] / (Le.Cd ? Ci)

With this the irrigation and drainage requirements for salinity control can can be computed too.

In irrigation projects in (semi)arid zones and climates it is important to check the leaching requirement, whereby the field irrigation efficiency (indicating the fraction of irrigation water percolating to the underground) is to be taken into account.

The desired soil salinity level Cd depends on the crop tolerance to salt. The University of Wyoming, USA, and the Government of Alberta, Canada, report crop tolerance data.

Strip cropping: an alternative

In irrigated lands with scarce water resources suffering from drainage (high water table) and soil salinity problems, strip cropping is sometimes practiced with strips of land where every other strip is irrigated while the strips in between are left permanently fallow.

Owing to the water application in the irrigated strips they have a higher watertable which induces flow of groundwater to the unirrigated strips. This flow functions as subsurface drainage for the irrigated strips, whereby the water table is maintained at a not-too-shallow depth, leaching of the soil is possible, and the soil salinity can be controlled at an acceptably low level.

In the unirrigated (sacrificial) strips the soil is dry and the groundwater comes up by capillary rise and evaporates leaving the salts behind, so that here the soil salinizes. Nevertheless, they can have some use for livestock, sowing salinity resistant grasses or weeds. Moreover, useful salt resistant trees can be planted like Casuarina, Eucalyptus or Atriplex, keeping in mind that the trees have deep rooting systems and the salinity of the wet subsoil is less than of the topsoil. In these ways wind erosion can be controlled. The unirrigated strips can also be used for salt harvesting.

The majority of the computer models available for water and solute transport in the soil (e.g. SWAP, DrainMod-S, UnSatChem), are based on Richard's differential equation for the movement of water in unsaturated soil in combination with a differential salinity dispersion equation.

The models require input of soil characteristics like the relation between unsaturated soil moisture content, water tension, hydraulic conductivity and dispersivity. These relations vary to a great extent from place to place and are not easy to measure. The models use short time steps and need at least a daily data base of hydrological phenomena. Altogether this makes model application to a fairly large project the job of a team of specialists with ample facilities.

Simpler models, like SaltMod, based on seasonal water and soil balances and an empirical capillary rise function, are also available. They are useful for long-term salinity predictions in relation to irrigation and drainage practices.Spatial variations owing to variations in topography can be simulated and predicted using salinity cum groundwater models, like SahysMod.

WATER RESOURCES

Water resources are sources of water that are useful or potentially useful to humans. Uses of water include agricultural, industrial, household, recreational and environmental activities. Virtually all of these human uses require fresh water.

97% of water on the Earth is salt water, and only 3% is fresh water of which slightly over two thirds is frozen in glaciers and polar ice caps. The remaining unfrozen freshwater is mainly found as groundwater, with only a small fraction present above ground or in the air.

Fresh water is a renewable resource, yet the world's supply of clean, fresh water is steadily decreasing. Water demand already exceeds supply in many parts of the world and as the world population continues to rise, so too does the water demand. Awareness of the global importance of preserving water for ecosystem services has only recently emerged as, during the 20th century, more than half the world's wetlands have been lost along with their valuable environmental services. Biodiversity-rich freshwater ecosystems are currently declining faster than marine or land ecosystems. The framework for allocating water resources to water users (where such a framework exists) is known as water rights.

Sources of fresh water

Surface water is water in a river, lake or fresh water wetland. Surface water is naturally replenished by precipitation and naturally lost through discharge to the oceans, evaporation, evapotranspiration and sub-surface seepage.

Although the only natural input to any surface water system is precipitation within its watershed, the total quantity of water in that system at any given time is also dependent on many other factors. These factors include storage capacity in lakes, wetlands and artificial reservoirs, the permeability of the soil beneath these storage bodies, the runoff characteristics of the land in the watershed, the timing of the precipitation and local evaporation rates. All of these factors also affect the proportions of water lost.

Human activities can have a large and sometimes devastating impact on these factors. Humans often increase storage capacity by constructing reservoirs and decrease it by draining wetlands. Humans often increase runoff quantities and velocities by paving areas and channelizing stream flow.

The total quantity of water available at any given time is an important consideration. Some human water users have an intermittent need for water. For example, many farms require large quantities of water in the spring, and no water at all in the winter. To supply such a farm with water, a surface water system may require a large storage capacity to collect water throughout the year and release it in a short period of time. Other users have a continuous need for water, such as a power plant that requires water for cooling. To supply such a power plant with water, a surface water system only needs enough storage capacity to fill in when average stream flow is below the power plant's need.

Nevertheless, over the long term the average rate of precipitation within a watershed is the upper bound for average consumption of natural surface water from that watershed.

Natural surface water can be augmented by importing surface water from another watershed through a canal or pipeline. It can also be artificially augmented from any of the other sources listed here, however in practice the quantities are negligible. Humans can also cause surface water to be "lost" (i.e. become unusable) through pollution.

Brazil is the country estimated to have the largest supply of fresh water in the world, followed by Russia and Canada.

Under river flow

Throughout the course of the river, the total volume of water transported downstream will often be a combination of the visible free water flow together with a substantial contribution flowing through sub-surface rocks and gravels that underlie the river and its floodplain called the hyporheic zone. For many rivers in large valleys, this unseen component of flow may greatly exceed the visible flow.

The hyporheic zone often forms a dynamic interface between surface water and true ground-water receiving water from the ground water when aquifers are fully charged and contributing water to ground-water when ground waters are depleted. This is especially significant in karst areas where pot-holes and underground rivers are common.

Ground water

Sub-surface water, or groundwater, is fresh water located in the pore space of soil and rocks. It is also water that is flowing within aquifers below the water table. Sometimes it is useful to make a distinction between sub-surface water that is closely associated with surface water and deep sub-surface water in an aquifer (sometimes called "fossil water"). Sub-surface water can be thought of in the same terms as surface water: inputs, outputs and storage. The critical difference is that due to its slow rate of turnover, sub-surface water storage is generally much larger compared to inputs than it is for surface water. This difference makes it easy for humans to use sub-surface water unsustainably for a long time without severe consequences. Nevertheless, over the long term the average rate of seepage above a sub-surface water source is the upper bound for average consumption of water from that source.

The natural input to sub-surface water is seepage from surface water. The natural outputs from sub-surface water are springs and seepage to the oceans. If the surface water source is also subject to substantial evaporation, a sub-surface water source may become saline. This situation can occur naturally under endorheic bodies of water, or artificially under irrigated farmland.

In coastal areas, human use of a sub-surface water source may cause the direction of seepage to ocean to reverse which can also cause soil salinization. Humans can also cause sub-surface water to be "lost" (i.e. become unusable) through pollution. Humans can increase the input to a sub-surface water source by building reservoirs or detention ponds.

Desalination

Desalination is an artificial process by which saline water (generally sea water) is converted to fresh water. The most common desalination processes are distillation and reverse osmosis. Desalination is currently expensive compared to most alternative sources of water, and only a very small fraction of total human use is satisfied by desalination. It is only economically practical for high-valued uses (such as household and industrial uses) in arid areas. The most extensive use is in the Persian Gulf.

Frozen water

Several schemes have been proposed to make use of icebergs as a water source, however to date this has only been done for novelty purposes. Glacier runoff is considered to be surface water.

The Himalayas, which are often called "The Roof of the World", contain some of the most extensive and rough high altitude areas on Earth as well as the greatest area of glaciers and permafrost outside of the poles. Ten of Asia's largest rivers flow from there, and more than a billion people's livelihoods depend on them. To complicate matters, temperatures are rising more rapidly here than the global average.

In Nepal the temperature has risen with 0.6 degree over the last decade, whereas the global warming has been around 0.7 over the last hundred years. Uses of fresh water can be categorized as consumptive and non-consumptive (sometimes called "renewable"). A use of water is consumptive if that water is not immediately available for another use. Losses to sub-surface seepage and evaporation are considered consumptive, as is water incorporated into a product (such as farm produce). Water that can be treated and returned as surface water, such as sewage, is generally considered non-consumptive if that water can be put to additional use.

Agricultural

It is estimated that 69% of worldwide water use is for irrigation, with 15-35% of irrigation withdrawals being unsustainable. It takes around 3,000 litres of water, converted from liquid to vapour, to produce enough food to satisfy one person's daily dietary need. This is a considerable amount, when compared to that required for drinking, which is between two and five litres. To produce food for the 6.5 billion or so people who inhabit the planet today requires the water that would fill a canal ten metres deep, 100 metres wide and 7.1 million kilometres long - that's enough to circle the globe 180 times.

Increasing water scarcity

Fifty years ago, the common perception was that water was an infinite resource. At this time, there were fewer than half the current number of people on the planet. People were not as wealthy as today, consumed fewer calories and ate less meat, so less water was needed to produce their food. They required a third of the volume of water we presently take from rivers. Today, the competition for water resources is much more intense.

This is because there are now nearly seven billion people on the planet, their consumption of water-thirsty meat and vegetables is rising, and there is increasing competition for water from industry, urbanisation and biofuel crops. In future, even more water will be needed to produce food because the Earth's population is forecast to rise to 9 billion by 2050. An additional 2.5 or 3 billion people, choosing to eat fewer cereals and more meat and vegetables could add an additional five million kilometres to the virtual canal mentioned above.

An assessment of water management in agriculture was conducted in 2007 by the International Water Management Institute in Sri Lanka to see if the world had sufficient water to provide food for its growing population. It assessed the current availability of water for agriculture on a global scale and mapped out locations suffering from water scarcity.

It found that a fifth of the world's people, more than 1.2 billion, live in areas of physical water scarcity, where there is not enough water to meet all demands. A further 1.6 billion people live in areas experiencing economic water scarcity, where the lack of investment in water or insufficient human capacity make it impossible for authorities to satisfy the demand for water.

The report found that it would be possible to produce the food required in future, but that continuation of today's food production and environmental trends would lead to crises in many parts of the world. To avoid a global water crisis, farmers will have to strive to increase productivity to meet growing demands for food, while industry and cities find ways to use water more efficiently.

In some areas of the world irrigation is necessary to grow any crop at all, in other areas it permits more profitable crops to be grown or enhances crop yield. Various irrigation methods involve different trade-offs between crop yield, water consumption and capital cost of equipment and structures. Irrigation methods such as furrow and overhead sprinkler irrigation are usually less expensive but are also typically less efficient, because much of the water evaporates, runs off or drains below the root zone.

Other irrigation methods considered to be more efficient include drip or trickle irrigation, surge irrigation, and some types of sprinkler systems where the sprinklers are operated near ground level. These types of systems, while more expensive, usually offer greater potential to minimize runoff, drainage and evaporation. Any system that is improperly managed can be wasteful, all methods have the potential for high efficiencies under suitable conditions, appropriate irrigation timing and management. Some issues that are often insufficiently considered are salinization of sub-surface water and contaminant accumulation leading to water quality declines.

As global populations grow, and as demand for food increases in a world with a fixed water supply, there are efforts under way to learn how to produce more food with less water, through improvements in irrigation methods and technologies, agricultural water management, crop types, and water monitoring. Aquaculture is a small but growing agricultural use of water. Freshwater commercial fisheries may also be considered as agricultural uses of water, but have generally been assigned a lower priority than irrigation (see Aral Sea and Pyramid Lake).

Industrial

It is estimated that 22% of worldwide water use is industrial. Major industrial users include power plants, which use water for cooling or as a power source (i.e. hydroelectric plants), ore and oil refineries, which use water in chemical processes, and manufacturing plants, which use water as a solvent. The portion of industrial water usage that is consumptive varies widely, but as a whole is lower than agricultural use. Water is used in power generation. Hydroelectricity is electricity obtained from hydropower. Hydroelectric power comes from water driving a water turbine connected to a generator. Hydroelectricity is a low-cost, non-polluting, renewable energy source. The energy is supplied by the sun. Heat from the sun evaporates water, which condenses as rain in higher altitudes, from where it flows down.

Pressurized water is used in water blasting and water jet cutters. Also, very high pressure water guns are used for precise cutting. It works very well, is relatively safe, and is not harmful to the environment. It is also used in the cooling of machinery to prevent over-heating, or prevent saw blades from over-heating.

Water is also used in many industrial processes and machines, such as the steam turbine and heat exchanger, in addition to its use as a chemical solvent. Discharge of untreated water from industrial uses is pollution. Pollution includes discharged solutes (chemical pollution) and discharged coolant water (thermal pollution). Industry requires pure water for many applications and utilizes a variety of purification techniques both in water supply and discharge.

CANAL IRRIGATION STATE-WISE IN INDIA

Canals used to be the most important source of irrigation up-to 1960s, but in the 1970s they yielded first place to wells and tube wells and now constitute the second most important source of irrigation in India.

The percentage of canal irrigation area to total irrigated area in the country has fallen from about 39.77 per cent in 1950-51 to 29 per cent in 2000-01.

Canals can be an effective source of irrigation in areas of low level relief, deep fertile soils, perrenial source of water and extensive command area. Therefore, the main concentration of canal irrigation is in the northern plain of India, especially the areas comprising Uttar Pradesh Haryana and Punjab. The digging of canals in rocky and uneven areas is difficult and uneconomic. Thus the canals are practically absent from the Peninsular plateau area. However, the coastal and the delta regions in South India do have some canals for irrigation.

Broadly speaking, canals in India are of two types, viz., (i) inundation canals, which are taken out from the rivers without any regulating system like weirs etc. at their head. Such canals provide irrigation mainly in the rainy season when the river is in flood and there is excess water.

When the rainy season is over, the flood in the river subsides, the level of water falls below the level of the canal head and the canal dries up. Some canals taken off from the Satluj in Punjab were of this type. Since irrigation from this type of canals is uncertain, they have been converted in perennial canals. (ii) Perennial Canals are those which are taken off from perennial rivers by constructing a barrage across the river. Most of the canals in India today are perennial.

The net area under canal irrigation is about 15.8 million hectares. The main canal irrigated areas are in the northern plains of India where Uttar Pradesh, Punjab, Haryana, Rajasthan and Bihar account for about 60 per cent of the canal irrigated area of the country. In south and central India, Andhra Pradesh, Maharasthra, Karnataka, Madhya Pradesh, Chhattisgarh, Orissa and Tamil Nadu are important states of canal irrigation.

Uttar Pradesh:

Canals constitute an important source of irrigation in Uttar Pradesh. The state is drained by perennial rivers originating in the snow covered Himalayan ranges and is blessed with fertile soils. But the amount of rainfall, especially in western parts of the state, is not sufficient for sustained agricultural growth.

Therefore, a large number of canals have been constructed to provide regular supply of sufficient water to the crops. Uttar Pradesh has about 3,091 thousand hectares under canal irrigation which is 30.91 per cent of the total canal irrigated area of the country. Over one-fourth of the net irrigated area of the state is irrigated by canals. Following are the main canals.

1. Upper Ganga Canal:

This canal takes off from the Ganga at Kankhal (Haridwar). The construction of this canal commenced in 1842 and completed in 1854. The main canal is 342 km long while the length of its distributaries is about 6,200 km. During the first 32 km of its course, between Haridwar and Roorkee, it passes through a broken country so that at some places it is taken over the rivers and at others below the rivers. It irrigates about 7 lakh hectares of land in the upper part of the Ganga-Yamuna Doab. Districts of Saharanpur, Meerut, Ghaziabad, Bulandshahar, Aligarh, Mathura, Etah, Kanpur, Mainpuri, Farrukhabad and Fatehpur get benefit from this canal. Its main branches are Devbandh, Anupshahar, Motta and Hathras. It joins with the Lower Ganga Canal at Mainpuri and the water in this canal is considerably increased. Further beyond, these two canals flow separately.

2. Lower Ganga Canal:

This canal was taken from the Ganga near Narora (Bulandshahar) in 1878. The length of the canal including its distributaries is about 6000 km. Its main branches are Etawah, Kanpur and Fatehpur. It irrigates about 4.6 lakh hectares in the districts of Bulandshahar, Farrukhabad, Mainpuri, Aligarh, Etah, Etawah, Fatehpur, Kanpur and Allahabad.

3. Sharda Canal:

As its name indicates, this canal is taken from the Sharda River at Banbasa near the Indo-Nepal border. The construction work on this canal was completed in 1928. The total length of the main canal and its distributaries is 13,624 km. It is thus one of the longest canal systems of the world. This canal system irrigates about eight lakh hectares of land mainly in Allahabad, Sultanpur, Pilibhit, Bareilly, Hardoi, Shahjahanpur, Sitapur, Lucknow, Barabanki, Rai Bareli, Unnao, Partapgarh and Kheri, districts.

4. Eastern Yamuna Canal:

It takes off from the river Yamuna at Faizabad. It was constructed in 1831. The main canal and its distributaries cover a distance of 1,450 km and irrigate about 2 lakh hectares of land in the districts of Saharanpur, Muzaffarnagar, Meerut and Ghaziabad. It again joins the Yamuna River at Delhi and irrigates a part of the union territory also.

5. Agra Canal:

This canal is taken from the right bank of the Yamuna at Okhla (Delhi). It was built in 1874 and irrigates about 1.5 lakh hectares in Agra and Mathura in U.P., Faridabad in Haryana, Bharatpur in Ra jasthan and also parts of union territory of Delhi.

6. Betwa Canal:

Built in the Third Five Year Plan, this canal takes off from the Betwa River about 56 km south-west of Jhansi. It irrigates about 1.2 lakh hectares in Jhansi, Jalaun and Hamirpur districts. Apart from the above mentioned major canals, some other canals such as Ken, Chambal, Dhasan and Son canal also irrigate some areas in the southern part of Uttar Pradesh.

Punjab:

In early days, canal irrigation accounted for about 39-14 per cent of the total irrigated area in Punjab but the share of canal irrigation fell down to less than 19 per cent in 2000-01. Following are some of the important canals of Punjab.

1. Upper Bari Doab Canal:

This canal is taken from the Ravi River at Madhopur near Pathankot. Construction of this canal started in 1849 and completed in 1859. It irrigates about 3 lakh hectares in Gurdaspur and Amritsar districts.

2. Sirhind Canal:

This canal was taken from the Satluj River at Rupnagar (Ropar) in the year 1886-87. The total length of the canal along with its distributaries is 6,115 km. Its main branches are the Patiala, Abohar, Bhatinda, Kotla and Ghaggar. It irrigates about 7 lakh hectares in Patiala, Sangrur, Bhatinda, Ludhiana and Ferozepur districts.

The Kotla and the Ghaggar branches provide irrigation to Hissar, Sirsa and Fatehabad districts of Haryana also. In order to augment the supply of water, the Sirhind Feeder Canal was completed in 1960. It takes off from the Ferozepur Feeder at its 18th km at Malanwala.

It is 142 km long and supplies water to Abohar branch of the Sirhind Canal. This water is drawn from the Satluj and Beas rivers which used to go unused to Pakistan. It also provides irrigation to Ferozepur, Faridkot and Muktsar districts in Punjab and to some parts of Rajasthan.

3. Bhakra Canal:

It draws water from the Bhakra dam built across the Satluj River. It was completed in 1954. It irrigates a vast area of about 15 lakh hectares in Punjab, Haryana and Rajasthan. In Punjab, Ludhiana, Patiala, Sangrur, Jalandhar, and Ferozepur districts are benefited by this canal.

4. Bist Doab Canal:

It is a part of the Bhakra-Nangal Project. The total length of this canal along with its distributaries is 1,090 km. It irrigates about 4 lakh hectares in Jalandhar and Hoshiarpur districts.

Bikaner and Beas are other projects which provide irrigation to parts of Punjab and the neighbouring states of Haryana and Rajasthan.

Haryana:

Haryana depends upon canal irrigation for its agricultural prosperity to a great extent. About 49-89 per cent of the irrigated area in Haryana is irrigated by canals. This is the second highest percentage for any state after Chhattisgarh. Following are the main canals.

1. The Western Yamuna Canal:

It takes off from the right bank of the Yamuna at Tajewala. It was built by Feroze Shah Tughlak. The total length of the canal along with its distributaries is 3,200 km and it irrigates about 4 lakh hectares in Ambala, Kurukshetra, Kamal, Panipat, Rohtak, Hissar, Sirsa, Faridabad and Jind districts. It's important branches are, the Delhi, the Hansi and the Sirsa branch.

2. Bhakra Canal:

After irrigating Punjab areas, the Bhakra canal enters Haryana near Tohana and irrigates large parts of Hissar, Fatehabad and Sirsa districts. Its main branches are the Fatehabad, the Ratia, the Rori, the Barwala and the Tohana branch.

3. Jui Canal:

This is a lift irrigation scheme designed to irrigate the semi-desert region of Bhiwani and adjoining areas. This 169 km long canal irrigates about 32 thousand hectares.

4. Gurgaon Canal:

It takes off from the Yamuna at Okhla in Delhi. Started in 1970, this canal irrigates 1.2 lakh hectares in Gurgaon and Faridabad districts.

Andhra Pradesh:

With 1,649 thousand hectares under canal irrigation i.e. 10.31% of the total canal irrigated area of India, Andhra Pradesh is next only to Uttar Pradesh and tops in South India in so far as area under canal irrigation is concerned. Canal irrigation accounts for about 36-42 per cent of the net irrigated area of the state. The major canals of Andhra Pradesh are taken off from the Krishna, the Godavari and the Tungabhadra rivers and the major canal irrigated areas are in the deltas and the coastal regions.

The Godavari delta project comprises of two weirs — the Dowlaiswaram and the Ralli, which were completed in 1846. From these, right bank canal and the delta canal have been taken to irrigate about 4.5 lakh hectares. The Krishna delta project comprises Vijayawada anicut. Sunkesula anicut across the Krishna and the Tungabhadra irrigates about 4.5 lakh hectares.

The Kurnool-Cuddapah canal was taken off from the Tungabhadra in 1816. It irrigates about 1.2 lakh hectares in Kumool and Cuddappah districts. The Nagarjunasagar Project consists of a dam across the Krishna River in Nalgonda district from where two canals have been taken off.

The right bank canal is 204 km long and irrigates 6.7 lakh hectares in Guntur, Nalgonda, Kumool and Krishna districts. The left bank canal is 171 km long and irrigates 3.56 lakh hectares in Khammam, Krishna and West Godavari districts.

Bihar:

Rainfall in several parts of Bihar is inadequate, unreliable and uncertain leading to severe and frequent droughts. This has made canal irrigation an important part of agricultural practice in Bihar. Canal irrigation accounts for over 31 per cent of the total area of the state.

1. Sone Canals:

The Eastern Sone Canal was taken from the Sone River at Varun in 1857. This 130 km long canal irrigates 2-5 lakh hectares in Patna and Gaya clistricts. The Western Sone Canal has been taken from this river at Dehri. It provides irrigation to Shahabad district.

2. Kosi Canals:

Two canals known as the Eastern and the Western Kosi Canal have been taken * from the eastern and the western banks of the river respectively by constructing a multipurpose dam near the Indo-Nepal border. These two canals have the irrigation potential of 4-5 and 3-5 lakh hectares respectively. Pumea, Muzaffarpur, Darbhanga, Champaran and Saran districts are benefited by this project. Besides, 4 lakh hectares are irrigated in Nepal.

3. Gandak Canals:

A 743 metre long dam has been built across the Gandak at Balmiki Nagar from where two canals have been taken off. The eastern canal is called Tirhut. It is 250 km long and irrigates about 7 lakh hectares in Champaran, Darbhanga, and Muzaffarpur districts.

The western canal is known as Saran Canal and irrigates 7-6 hectares of Saran district of Bihar and Deoria and Gorakhpur districts of Uttar Pradesh. Another Triveni Canal has been taken from Triveni to irrigate about one lakh hectares of Champaran district.

West Bengal:

Although most parts of West Bengal receive sufficient rainfall and do not require irrigation, still some parts of the state do feel the necessity of irrigation. More than one third of the net irrigated area is irrigated by canals. The Mayurakshi project was completed in 1951 at Tilpara.

Two canals have been taken from this project to irrigate large areas in Bir Bhum, Murshidabad and Bardhaman districts. The main canal from the Kaugsabati completed in 1965 irrigates 1.2 lakh hectares in Bankura and Midnapur districts.

The canals from the Damodar Valley Corporation irrigate over five lakh hectares in West Bengal. The 520 km long Midnapur Canal has been taken off from the Kosi at Midnapur and irrigates about ten thousand hectares.

Rajasthan:

Most of Rajasthan is a desert area and largely depends upon irrigation for successful growth of crops. Canals form an important section of irrigation and account for about 27.6 per cent of the net irrigated area of the state.

1. Indira Gandhi Canal:

This canal originates from Harike Barrage near the confluence of Satluj and Beas rivers in Ferozepur district of Punjab. The plan for this canal was prepared in 1957-58 and the work on this project started on 31st March 1958. The canal does not do any irrigation in Punjab and is known as Rajasthan Feeder.

The total length of Rajasthan Feeder is 204 km. The main canal is 40 metres wide at bottom and 6.4 metres deep. The carrying capacity of canal is 18,500 cusecs of water at its head. According to a proposal in 1981, Rajasthan was allocated 8.6 million acre feet of Ravi-Beas surplus water. The Indira Gandhi Canal envisages the utilisation of 7.6 million acre feet of water allocated to Rajasthan.

The head of the main canal is located near Masitanwali in Hanumangarh district. The tail of the 445 km long main canal is located near Mohangarh in Jaisalmer district. The Command Area of the canal is further extended till Gadra Road in Barmer district, through Sagarmal Gopa branch.

Construction work of the project is being carried out in two stages. Water was released in the main canal on 11 October 1967 and reached its tail on January 1987. The main canal has two kinds of branches and distributaries. The right bank branches of the canal are flow channels as the land west of the main channels slopes down gently towards Pakistan border.

The left bank branches except Rawatsar Branch which takes off from the head of the main canal, are lift channels. This is because of the fact that area towards the south-east of the main canal slopes towards the canal and water is to be lifted against the slope of the land.

Stage I:

Construction work of Stage I has been completed at a cost of Rs. 246 crore. It included the construction of 204 kms long feeder, 189 kms of the main canal and 2,960 kms long distribution system. Stage I has five flow branches and one lift canal covering south and south-western part of Ganganagar district and north and north-western parts of Bikaner district.

Nearly 4.79 lakh hectares of land is provided with flow irrigation and 0.46 lakh hectares get lift irrigation. The irrigation potential on full development will be 5.78 lakh hectares. The project plan of Stage I envisages intensive irrigation with an irrigation intensity of 110%. Irrigation intensity is expressed as percentage ratio between gross irrigated area and culturable command area of the project.

The above facilities will lead to annual food production of 14.50 lakh tonnes.

Stage II:

The construction work on Stage II is still in progress although it was to be completed in 1978 according to its original plan. At the current rates, it will cost Rs. 1,420 crore. Stage II of the project includes the construction of 256 km long main canal and 4,800 km long distributaries. According to the revised plan, this stage will provide extensive irrigation.

Extensive irrigation means reducing per acre allowance of water and providing irrigation to maximum cultivated area. Irrigation intensity of this region is 80% that amounts to provide irrigation to 80% of cultivable command area. This will prevent water-logging and soil-salinity and help in growing light irrigated crops.

Stage II proposes to develop irrigated pastures on an area of about 3.66 lakh hectares. This would help in providing benefits of irrigation to aboriginal nomadic communities, develop animal husbandry and arrest desertification. This stage of the project envisages to provide flow irrigation to seven lakh hectares and lift irrigation to 3.12 lakh hectares of land. This will lead to annual food production of 22.50 lakh tonnes.

2. Chambal Project:

This is a joint venture of Rajasthan and Madhya Pradesh. Under this project, Gandhi Sagar Dam has been constructed. Canals taken off from this dam irrigate about 5.15 lakh hectares in Rajasthan and Madhya Pradesh. In the second stage, Rana Partap Dam has been constructed which provides irrigation to 1.2 lakh hectares. In the third stage Jawahar Sagar Dam has been constructed.

3. The Gang or Bikaner Canal:

It was taken off from the river Satluj at Husainwala in 1928. This canal is a blood transfusion from the living Punjab into the moribund marusthal. It provides irrigation to about 3.4 lakh hectares in Bikaner division. The total length of this canal system is 1,280 km.

The other canal irrigation projects of Rajasthan include Jawai project on the Jawai river, Parbati project near Dhaulpur, Gudha project in Bundi District, Ghaggar canal, Pichuna Canal, Banas Canal, Bharatpur Canal and Urmila Canal. The Bhakra Canal also irrigates about 2.3 lakh hectares in Rajasthan.

Madhya Pradesh and Chhattisgarh:

The unevem surface, rocky structure and limited fertile soil have not encouraged the construction of canals and canals irrigation is not much important, keeping in view the size of these two states. However, a few canal systems have come up.

Consequently 68.9 per cent of the net irrigated area in Chhattisgarh and 19.49 per cent of the net irrigated area in Madhya Pradesh is under canal irrigation. Most of the canal irrigated area is in Gwalior, Bhind and Morena districts. The Chambal project provides irrigation to about 2.83 lakh hectares in these districts. The Weinganga canal takes off from the Weinganga River and irrigates 4,000 hectares of land in Balaghat and Seoni districts. Besides it provides irrigation to Bhandara district of Maharashtra. The Barna project involves the construction of a barrage across the Barna River (a tributary of the Narmada).

Canal taken out from this barrage irrigate about 64,400 hectares of land in Raisen district. The Tawa project canals originate from the barrage constructed across the Tawa River, and irrigate about 3 lakh hectares of land in Hoshangabad district. The Mahanadi canal takes off from Mahanadi River at Rudri in Raipur district. A subsidiary reservoir has been built at Maramsilli.

The length of the main canal and distributaries is 215 km and 1,175 km respectively. It irrigates about 3 lakh hectares. The Tandula canal takes off from the confluence of the Tandula and Sukha rivers by building two separate earthen dams. It irrigates about 1.1 lakh hectares in Raipur and Durg districts.

Orissa:

Canals irrigate about 878 thousand hectares of land which is nearly 25% of the net irrigated area of the state. Canals taken off from the Hirakud dam on the Mahanadi form a major irrigation network and provide irrigation to about 2.4 lakh hectares in Bolangir and Sambalpur districts. The 3,650 km long canal system in the Mahanadi delta region provides irrigation to about 4 lakh hectares in Cuttack and Puri districts. Taldanda Canal irrigates 62 thousand hectares in Mahanadi catchment area.

Karnataka:

Over 36.5 per cent of the net irrigated area in Karnataka is irrigated by canals. The Ghatprabha valley scheme developed on the Ghatprabha River is the most important irrigation project and irrigates about 3.2 lakh hectares in Belgaum and Bijapur districts. Canals of the Tungbhadra project irrigate about 2.7 lakh hectares in Bellary and Raichur districts.

The canals of Malprabha project irrigate about one lakh hectares in Belgaum, Dharwar and Bijapur districts and those of Bhadra Project also irrigate one lakh hectares in Shimoga district. The Visveswaraya Canal taken off from the Krishndraja Sagar Dam at the Cauvery irrigates about 50 thousand hectares in Mysore and Mandya districts.

Tamil Nadu:

Tamil Nadu has about 29 per cent of its net irrigated area under canal irrigation. The most important canal system lies in the Cauvery delta where 6,400 km long canals irrigate about 4 lakh hectares in Thanjavur and Tiruchchirapalli districts.

The Mettur Canal system of the Mettur dam on the Cauvery River irrigates about 1.2 lakh hectares in Salem and Tiruchchirapalli districts. The Lower Bhawani Project canal system irrigates about 79 thousand hectares in Coimbatore district. Parambikulam Aliyar and Manimuthar are other projects.

Maharashtra:

Over 35 per cent of the net irrigated area in Maharashtra is irrigated by canals. Over 3,000 km long canals irrigate over 4 lakh hectares. The right bank and the left bank canals of the Mutha Project across Mutha River (a tributary of the Bhima River) at Khadakvasla irrigate about 45 thousand hectares in Pune district besides providing potable water to Pune and Kirkee.

The canals of the Nira project irrigateover 66 thousand hectares in Pune, Satara and Solapur districts. Two canals taken out of the Godavari a Darana dam irrigate about 27 thousand hectares in Nashik and Ahmednagar districts Canals taken out from Gangapur dam at Godavari irrigate 33 thousand hectares in Nashik district. The take off from Bhandardara masonary dam at Pravara River. Both the left bank and the right bank canals irrigate about 34 thousand hectares in Ahmednagar.

The Tapi canal system is a joint venture of Maharashtra and Gujarat and irrigates about 3 lakh hectares. The other irrigation canals are those of Mula, Vir,

Puma, Gima, Jayakawadi, Wama, etc. Some other projects are at different stages of their completion.

Gujarat:

Mahi project stage I and II is designed to irrigate over 2 lakh hectares in Kheda and Panchmahals districts Canals of Ukai project on the Tapi river irrigate over one lakh hectares. The other important irrigation projects of Gujarat are Rudramala, Ozat, Dantiwada, Panam and Kaprapara.

Merits of Canal Irrigation:

1. Most of the canals provide perennial irrigation and supply water as and when needed. This saves the crops from drought conditions and helps in increasing the farm production.
2. Canals carry a lot of sediment brought down by the rivers. This sediment is deposited in the agricultural fields which add to the fertility of soil.
3. Some of the canals are parts of multipurpose projects and, therefore, provide cheap source of irrigation.
4. Although the initial cost involved in canal irrigation is much higher, it is quite cheap in the long run.

Demerits of Canal Irrigation

1. The canal water soaks into the ground and leads to the problem of water-logging along the canal route.
2. Excessive flow of water in the fields raises the ground water level. Capillary action brings alkaline salts to the surface and makes large areas unfit for agriculture. Vast areas in Panjab, Haryana and Uttar Pradesh suffer from the problem of 'reh' caused by canal irrigation. About 36,000 hectares have been rendered useless in Nira Valley of Maharashtra due to high concentration of salts in the soil resulting from canal irrigation.

11

SUITABILITY OF WATER FOR IRRIGATION

The suitability of irrigation water is mainly depends on the amounts and type of salts present in water. The main soluble constituents are calcium, magnesium, sodium as cations and chloride, sulphate, biocarbonate as anions. The other ions are present in minute quantities are boron, selenium, molybdenum and fluorine which are harmful to animals fed on plants grown with excess concentration of these ions.

Water scarcity is seen as a major constraint to intensify agriculture in a sustainable manner as an attempt to meet the food requirements of a rapidly growing human population. The ever increasing human population, climate change due to increased emissions of greenhouse gases (GHGs), and intensification of agriculture, are putting severe pressure on the world's two major non-renewable resources of soil and water, and thus pose a big challenge to produce sufficient food to meet the current food demand. The present world population of 7.3 billion people is predicted to grow to over 9 billion by 2050, with the majority of this population increase occurring in developing countries, most of which already face food shortages.

A 70% increase in current agricultural productivity will be required to produce sufficient food if these human population growth predictions prove to be correct. In this context, concerted efforts are being made globally to improve the effectiveness of water which will be used for enhancing the production of irrigated crops. Additionally, efforts are also being made to improve water harvesting and water conservation in rain-fed agriculture.

The concentration and composition of soluble salts in water will determine its quality for various purposes (human and livestock drinking, irrigation of crops, etc.). The quality of water is, thus, an important component with regard to sustainable use of water for irrigated agriculture, especially when salinity development is expected to be a problem in an irrigated agricultural area.

There are four basic criteria for evaluating water quality for irrigation purposes:

Total content of soluble salts (salinity hazard)

Relative proportion of sodium (Na+) to calcium (Ca2+) and magnesium (Mg2+) ions – sodium adsorption ratio (sodium hazard)

Residual sodium carbonates (RSC) – bicarbonate (HCO3-) and carbonate (CO32-) anions concentration, as it relates to Ca 2+ plus Mg2+ ions.

Excessive concentrations of elements that cause an ionic imbalance in plants or plant toxicity.

In order to achieve the first three important criteria, the following characteristics need to be determined in the irrigation waters: electrical conductivity (EC), soluble anions (CO32-, HCO3-, Cl- and SO42-) where Cl- and SO42- are optional and soluble cations (Na+, K+, Ca2+, Mg2+) where K is optional. Finally, boron level must also be measured. The pH of the irrigation water is not an acceptable criterion of water quality because the water pH tends to be buffered by the soil, and most crops can tolerate a wide pH range. A detailed description of the techniques commonly employed for the analysis of irrigation water is available (USSL Staff 1954; Bresler et al. 1982).

Soil scientists use the following categories to describe irrigation water effects on crop production and soil quality:

Salinity hazard – total soluble salt content

Sodium hazard – relative proportion of sodium to calcium and magnesium ions

pH – acid or basic

Alkalinity – carbonate and bicarbonate

Specific ions: chloride, sulfate, boron, and nitrate.

Another potential irrigation water quality impairment that may affect suitability for cropping systems is microbial pathogens.

SALINITY HAZARD

The main problem related to irrigation water quality is water salinity-which refers to the total amount of salts dissolved in the water, but it does not indicate which salts are present. High levels of salts in irrigation water reduce water availability to the crop (because of osmotic pressure) and cause yield reduction. Above a certain threshold, reduction in crop yield is proportional to the increase in salinity level. Different crops vary in their tolerance to salinity and, therefore, have different thresholds and yield reduction rates.

The most common parameters used for determining irrigation water quality in relation to its salinity are EC and TDS. If irrigation water salinity exceeds the threshold for the crop, yield reduction will occur.

TDS ppm or mg/L	EC dS/m	Salinity hazard
<500	<0.8	Low
500 - 1,000	0.8 - 1.6	Medium
1,000 - 2,000	1.6 - 3	High
> 2,000	> 3	Very high

The most influential water quality guideline on crop productivity is the water salinity hazard as measured by electrical conductivity (ECw). The primary effect of high ECw water on crop productivity is the inability of the plant to compete with ions in the soil solution for water (physiological drought). The higher the EC, the less water is available to plants, even though the soil may appear wet. Because plants can only transpire "pure" water, usable plant water in the soil solution decreases dramatically as EC increases. The amount of water transpired through a crop is directly related to yield; therefore, irrigation water with high ECw reduces yield potential (Table 2). Actual yield reductions from irrigating with high EC water varies substantially. Factors influencing yield reductions include soil type, drainage, salt type, irrigation system and management. Beyond effects on the immediate crop is the long term impact of salt loading through the irrigation water. Water with an ECw of only 1.15 dS/m contains approximately 2,000 pounds of salt for every acre foot of water. You can use conversion factors in Table 3 to make this calculation for other water EC levels.

TABLE 1. GENERAL GUIDELINES FOR SALINITY HAZARD OF IRRIGATION WATER BASED UPON CONDUCTIVITY.

Limitations for use	Electrical Conductivity
	(dS/m)*
None	?0.75

Some	0.76 - 1.5
Moderate	1.51 - 3.00
Severe	?3.00
*dS/m at 25°C = mmhos/cmLeaching required at higher range.Good drainage needed and sensitive plants may have difficulty at germination.	

TABLE 2. POTENTIAL YIELD REDUCTION FROM SALINE WATER FOR SELECTED IRRIGATED CROPS.

% yield reduction				
Crop	0%	10%	25%	50%
	ECw			
Barley	5.3	6.7	8.7	12
Wheat	4.0	4.9	6.4	8.7
Sugarbeet	4.7	5.8	7.5	10
Alfalfa	1.3	2.2	3.6	5.9
Potato	1.1	1.7	2.5	3.9
Corn (grain)	1.1	1.7	2.5	3.9
Corn (silage)	1.2	2.1	3.5	5.7
Onion	0.8	1.2	1.8	2.9
Dry Beans	0.7	1.0	1.5	2.4

Adapted from "Quality of Water for Irrigation." R.S. Ayers. Jour. of the Irrig. and Drain. Div., ASCE. Vol 103, No. IR2, June 1977, p. 140.

ECw = electrical conductivity of the irrigation water in dS/m at 25oC.

Sensitive during germination. ECw should not exceed 3 dS/m for garden beets and sugarbeets.

Other terms that laboratories and literature sources use to report salinity hazard are: salts, salinity, electrical conductivity (ECw), or total dissolved solids (TDS). These terms are all comparable and all quantify the amount of dissolved "salts" (or ions, charged particles) in a water sample. However, TDS is a direct measurement of dissolved ions and EC is an indirect measurement of ions by an electrode.

Although people frequently confuse the term "salinity" with common table salt or sodium chloride (NaCl), EC measures salinity from all the ions dissolved in a sample. This includes negatively charged ions (e.g., Cl, NO3) and positively charged ions (e.g., Ca, Na). Another common source of confusion is the variety of unit systems used with ECw. The preferred unit is deciSiemens per meter (dS/m), however millimhos per centimeter (mmho/cm) and micromhos per centimeter (µmho/cm) are still frequently used. Conversions to help you change between unit systems are provided in Table 3.

TABLE 3. CONVERSION FACTORS FOR IRRIGATION WATER QUALITY LABORATORY REPORTS.

Component	To Convert	Multiply By	To Obtain
Water nutrient or TDS	mg/L	1.0	ppm
Water salinity hazard	1 dS/m	1.0	1 mmho/cm
Water salinity hazard	1 mmho/cm	1,000	1 µmho/cm
Water salinity hazard	ECw (dS/m) for EC <5 dS/m	640	TDS (mg/L)
Water salinity hazard	ECw (dS/m) for EC >5 dS/m	800	TDS (mg/L)
Water NO3N, SO4-S,B applied	ppm	0.23	lb per acre inch
Definitions			
		Abbrev.	Meaning
		mg/L	milligrams per liter
		meq/L	milliequivalents per liter
		ppm	parts per million
		dS/m	deciSiemens per meter
		µS/cm	microSiemens per centimeter
		mmho/cm	millimhos per centimeter
		TDS	total dissolved solids
Irrigation water	acre inch	27,150	gallons of water

Dissolved salts

The natural weathering of rocks and soils results in salts being removed from a wide source and concentrated in rivers, dams and underground water resources. The dissolved salts in irrigation water are made up of positively (+) charged cations and negatively charged (-) anions charged. These molecules are attracted to each other and can be tightly or loosely bound to form salts.

The most common salts in irrigation water are table salt (sodium chloride), gypsum (calcium sulphate) epsom salts (magnesium sulphate), baking soda (sodium bicarbonate). The most common cations (+) are calcium, magnesium, and sodium while the most common anions are chloride, sulphates and bicarbonate. The ration of these ions vary from one water source to the other.

DEFINING SALINITY

The total salt concentration or total dissolved solids or TDS for short characterises the salinity or irrigation water. The test result will show you the TDS expressed in milligrams of slat per litre or mg/L or parts per million (ppm) per litre which is the same number as TDS. The higher this number the TDS the higher is the salinity of water. The other main measure of salinity is electrical conductivity or EC. This can be measured very easily by farm managers or irrigators in the field. This is a very technical subject and your test must be compared to various norms.

The water tests can be used together with your soil test to be able to identify the crop potential of a particular land under irrigation and monitor the build-up of salts in the soil over a long period. The solution may lie in applying more water to the land than can be held by the crop in the root zone so that excess salinity can be removed or leached from the soil profile. The proposed solutions must be carefully evaluated for possible financial gain and cost from a particular strategy and experts in this field must be consulted beforehand.

Crops and crop cultivars with differing salt tolerances can also be evaluated and those with a higher salt tolerance can be chosen for planting.

Sodium Hazard

The parameter used to determine the sodium hazard is SAR, or sodium adsorption ratio. This factor indicates the amount of sodium in the irrigation water in relation to calcium and magnesium. Calcium and magnesium tend to counter the negative effect of sodium.

High SAR levels might result in a breakdown of soil structure and water infiltration problems. Soil tends to seal and become hard and compact when it's too dry.

Ironically, higher salinity reduces the negative effect of sodium on soil structure. When sodium levels in the soil are high in relation to calcium and magnesium-in other words, when the SAR is high-flushing the soil with good-quality irrigation water will only exacerbate the problem.

Chloride

Chloride is a common ion in Colorado irrigation waters. Although chloride is essential to plants in very low amounts, it can cause toxicity to sensitive crops at high concentrations (Table 4). Like sodium, high chloride concentrations cause more problems when applied with sprinkler irrigation (Table 6). Leaf burn under sprinkler from both sodium and chloride can be reduced by night time irrigation or application on cool, cloudy days. Drop nozzles and drag hoses are also recommended when applying any saline irrigation water through a sprinkler system to avoid direct contact with leaf surfaces.

TABLE 4. CHLORIDE CLASSIFICATION OF IRRIGATION WATER.

Chloride (ppm)	Effect on Crops
Below 70	Generally safe for all plants.
70-140	Sensitive plants show injury.
141-350	Moderately tolerant plants show injury.
Above 350	Can cause severe problems.

Chloride tolerance of selected crops. Listing in order of increasing tolerance: (low tolerance) dry bean, onion, carrot, lettuce, pepper, corn, potato, alfalfa, sudangrass, zucchini squash, wheat, sorghum, sugar beet, barley (high tolerance). Source: Mass (1990) Crop Salt Tolerance. Agricultural Salinity Assessment and Management Manual. K.K. Tanji (ed.). ASCE, New York. pp 262-304.

Boron

Boron is another element that is essential in low amounts, but toxic at higher concentrations (Table 5). In fact, toxicity can occur on sensitive crops at concentrations less than 1.0 ppm. Colorado soils and irrigation waters contain enough B that additional B fertilizer is not required in most situations. Because B toxicity can occur at such low concentrations, an irrigation water analysis is advised for ground water before applying additional B to irrigated crops.

TABLE 5. BORON SENSITIVITY OF SELECTED COLORADO PLANTS (B CONCENTRATION, MG/ L*)

Sensitive		Moderately Sensitive	Moderately Tolerant	Tolerant
0.5-0.75	0.76-1.0	1.1-2.0	2.1-4.0	4.1-6.0
Peach	Wheat	Carrot	Lettuce	Alfalfa
Onion	Barley	Potato	Cabbage	Sugar beet
	Sunflower	Cucumber	Corn	Tomato
	Dry Bean		Oats	

Source: Mass (1987) Salt tolerance of plants. CRC Handbook of Plant Science in Agriculture. B.R. Cristie (ed.). CRC Press Inc.

*Maximum concentrations tolerated in soil water or saturation extract without yield or vegetative growth reductions. Maximum concentrations in the irrigation water are approximately equal to these values or slightly less.

Sulfate

The sulfate ion is a major contributor to salinity in many of Colorado irrigation waters. As with boron, sulfate in irrigation water has fertility benefits, and irrigation water in Colorado often has enough sulfate for maximum production for most crops. Exceptions are sandy fields with <1 percent organic matter and <10 ppm SO-S in irrigation water.

Nitrogen

Nitrogen in irrigation water (N) is largely a fertility issue, and nitrate-nitrogen (NO3-N) can be a significant N source in the South Platte, San Luis Valley, and parts of the Arkansas River basins. The nitrate ion often occurs at higher concentrations than ammonium in irrigation water. Waters high in N can cause quality problems in crops such as barley and sugar beets and excessive vegetative growth in some vegetables. However, these problems can usually be overcome by good fertilizer and irrigation management. Regardless of the crop, nitrate should be credited toward the fertilizer rate especially when the concentration exceeds 10 ppm NO3-N (45 ppm $NO3^-$). Table 3 provides conversions from ppm to pounds per acre inch.

WATER QUALITY PROBLEMS

Water used for irrigation can vary greatly in quality depending upon type and quantity of dissolved salts. Salts are present in irrigation water in relatively small but significant amounts. They originate from dissolution or weathering of the rocks and soil, including dissolution of lime, gypsum and other slowly dissolved soil minerals. These salts are carried with the water to wherever it is used. In the case of irrigation, the salts are applied with the water and remain behind in the soil as water evaporates or is used by the crop.

The suitability of a water for irrigation is determined not only by the total amount of salt present but also by the kind of salt. Various soil and cropping problems develop as the total salt content increases, and special management practices may be required to maintain acceptable crop yields. Water quality or suitability for use is judged on the potential severity of problems that can be expected to develop during long-term use.

The problems that result vary both in kind and degree, and are modified by soil, climate and crop, as well as by the skill and knowledge of the water user. As a result, there is no set limit on water quality; rather, its suitability for use is determined by the conditions of use which affect the accumulation of the water constituents and which may restrict crop yield. The soil problems most commonly encountered and used as a basis to evaluate water quality are those related to salinity, water infiltration rate, toxicity and a group of other miscellaneous problems.

Water Quality-Related Problems in Irrigated Agriculture

Salinity

Salts in soil or water reduce water availability to the crop to such an extent that yield is affected.

Water infiltration rate

Relatively high sodium or low calcium content of soil or water reduces the rate at which irrigation water enters soil to such an extent that sufficient water cannot be infiltrated to supply the crop adequately from one irrigation to the next.

Specific ion toxicity

Certain ions (sodium, chloride, or boron) from soil or water accumulate in a sensitive crop to concentrations high enough to cause crop damage and reduce yields.

Miscellaneous

Excessive nutrients reduce yield or quality; unsightly deposits on fruit or foliage reduce marketability; excessive corrosion of equipment increases maintenance and repairs.

Managing irrigation water quality problems

Infiltration problems resulting from low irrigation water quality

As mentioned earlier, SAR (sodium adsorption ratio) is an irrigation water parameter used to predict problems of water infiltration into soil. SAR is determined as:

Apart from water shortage-which is a result of water infiltration problems-some other related problems might occur as well; for example, weed growth, diseases, poor aeration, poor germination of seeds or root rot.

Various measures can be taken to overcome water infiltration problems that are related to water quality, including reducing the SAR of the water supply, cultivation and tillage, the addition of organic residues, irrigation management, and water or soil amendments.

Soil amendments and irrigation water quality

The purpose of soil amendments is to counter the effect of sodium by increasing the soluble calcium content or by increasing the salinity of the irrigation water.

Gypsum and other calcium-supplying materials

Gypsum is the most commonly used soil amendment. Since water infiltration problems caused by sodium affect mainly the upper few centimeters of soil, repeated small applications of gypsum-incorporated at lower rates into a shallow depth-are preferred over a single large application.

If the salinity of the irrigation water is low (EC<0.5 ds/m), gypsum can be added to the irrigation water at rates of one to four meq/l of dissolved calcium.

Other amendments

When lime ($CaCO_3$) is present in soil, some acids or acid-forming amendments can be used-these will cause calcium to be released into the soil solution. Some of these amendments are elemental sulfur, sulfuric acid and ferric sulfate.

Organic residues

These amendments improve soil structure and water infiltration by keeping the soil porous.

Blending irrigation water sources

Water infiltration can be improved either by increasing irrigation water salinity or by reducing the SAR. By diluting the irrigation water source with water having a lower sodium concentration, the SAR of the irrigation water is reduced, even if calcium and magnesium concentrations are higher.

Management of soil and irrigation water salinity

When salts build up in soil or in the growing medium, their concentration might become excessive. Salts are added to soil via irrigation water and with applied fertilizers. Applying more water than is needed by the crop leaches the salt below the root zone, deeper into the soil or out of the growing medium (when growing container plants). It is important to know how much to leach and when. The leaching requirement can be estimated from the following equation:

$$LR = ECw/ [5*ECe - ECw)]$$

Where LR is the minimum leaching requirement for the crop, ECw is the electrical conductivity of the irrigation water in ds/m (irrigation water salinity) and ECe is the soil EC tolerated by the crop, measured in a saturated soil extract.

The total irrigation water amount that has to be applied to meet both crop demand and your leaching requirement can be estimated from the equation:

$$AW = ET/ (1-LR)$$

Where AW is the amount of irrigation water that has to be applied, ET is the crop water demand and LR is the calculated leaching requirement. For example:

Crop water demand - 30m3/ha/day

ECe = 2.5 ds/m, ECw = 1.2 ds/m

LR = 1.2/(5*2.5-1.2) = 0.1

AW = 30/(1-0.1) = 33.33 m3/ha/day

IRRIGATION INTERVALS

Knowing the total amount of irrigation water to apply is not enough for managing salinity-irrigation intervals must also be considered. The appropriate irrigation intervals will depend not only on crop water demand, but also on factors such as the salinity threshold of the crop and the soil's capacity to hold water.

Applying the same amount of water to two soils with different characteristics will result in different wetting patterns and depths. Irrigation depth in a heavy soil is lower than in a sandy soil, since heavy soils hold more water than sandy soils-this means that heavy soils require higher application amounts of irrigation water at larger intervals in order to prevent an accumulation of salts exceeding the salinity threshold of the crop.

It must be noted, though, that salinity in the root zone increases between irrigations as a result of crop water uptake and water evaporation from the soil. The timing of leaching is not critical provided the salinity threshold of the crop is not exceeded.

IRRIGATION WATER QUALITY AND WASTEWATER RE-USE

Large-scale irrigation projects can bring prosperity to an area but less desirable changes can also occur as a result of increased intensity of land and water use. One important change in the hydrological regime is that of an alteration or degradation in quality that takes place as water is used and re-used within the hydrological basin. In addition, wastewater generated by agricultural and urban sources can degrade water quality and must be considered when developing a river basin management plan.

Agricultural subsurface drainage water presents the single greatest threat to water quality. The need for drainage is often quoted as a mechanism to eliminate the hazards from waterlogging and salinity in irrigated land. A drainage scheme can be implemented for engineering or economic reasons, but in either case the drainage water created by the scheme will contain a high concentration of salts. Careful consideration must be given to its disposal so that the water supplies downstream are not polluted.

The disposal of highly saline drainage water into river courses may need to be controlled in order to meet certain minimum standards of water quality for irrigated agriculture in downstream areas. Changes in downstream agricultural practices may be necessary to adapt to the inferior water quality, or alternative schemes may need to be implemented where the drainage or other wastewater is isolated form the main water supply. Due to the high cost of transporting wastewater to a disposal site (ocean, salt-sink or river discharge), the maximum number of uses of that water should be made before discharge. At that time, disposal must be in such a way that the river-basin water quality is protected and agricultural development is not jeopardized. All waste-water should be used and re-used until no longer fit for use.

Of equal importance when protecting the quality of water supplies that are to be used as a source of irrigation water is the utilization of effluent water from domestic sources or from an agricultural processing activity. Re-using wastewater can remove a potential cause of ground or surface water pollution and, at the same time, release higher quality water for other uses. Rising demands for good quality water for domestic and industrial uses in countries with highly developed economies have already created the necessity to re-use wastewater. Many developing countries are now facing a similar situation, especially in arid and semi-arid regions where limited water availability is already a severe constraint on development.

Agriculture is the major user of water and can accept lower quality water than domestic and industrial users. It is therefore inevitable that there will be a growing tendency to look toward irrigated agriculture for solutions to the overall effluent disposal problem. Because wastewater contains impurities, careful consideration must be given to the possible long-term effects on soils and plants from salinity, sodicity, nutrients and trace elements that occur normally manageable if associated problems with these impurities are understood and allowances made for them.

TABLE 6 EXISTING STANDARDS GOVERNING THE USE OF RENOVATED WATER IN AGRICULTURE

	California	Israel	South Africa	FR Germany
Orchards and vineyards	Primary[1] effluent; no spray irrigation; no use of dropped fruit	Secondary[2] effluent	Tertiary[3] effluent, heavily chlorinated where possible. No spray irrigation	No spray irrigation in the vicinity
Fodder fibre crops and seed crops	Primary effluent; surface or spray irrigation	Secondary effluent, but irrigation of seed crops for producing edible vegetables not permitted	Tertiary effluent	Pretreatment with screening and settling tanks. For spray irrigation, biological treatment and chlorination
Crops for human consumption that will be processed to kill pathogens	For surface irrigation, primary effluent. For spray irrigation, disinfected secondary effluent (no more than 23 coliform organisms per 100 ml)	Vegetables for human consumption not to be irrigated with renovated wastewater unless it has been properly disinfected (1000 coliform organisms per 100 ml in 80% of samples)	Tertiary effluent	Irrigation up to 4 weeks before harvesting only
Crops for human consumption in a raw state	For surface irrigation, no more than 2.2 coliform organisms per 100 ml. For spray irrigation, disinfected, filtered wastewater with turbidity of 10 units permitted, providing it has been treated by coagulation	Not to be irrigated with renovated wastewater unless they consist of fruits that are peeled before eating		Potatoes and cereals - irrigation through flowering stage only

Source: WHO (1973).

Primary treatment of wastewater refers to the settling and removal of a portion of the suspended organic and inorganic solids.

Secondary treatment refers to the activated sludge process and biological filtration(trickling filtration). It may also include retention.

Tertiary or Advanced Treatment includes several processes depending on the use of the final product but usually includes clarification, activated carbon treatment, denitrification and ion exchange.

The re-use of sewage effluent for agricultural practices is not an entirely new concept. Law (1968) cites 99 references on the use of sewage as an agricultural water resource. Some countries have developed standards for the use of effluents in terms of the treatment required and bacteriological characteristics, as presented in Table 6. A meeting of experts convened by WHO (1973) concluded that primary treatment would be sufficient to permit re-use for the irrigation of crops that are not for direct human consumption.

Secondary treatment and most probably disinfection and filtration are con

Source: WHO (1973).

Health criteria:

1. Freedom from gross solids; significant removal of parasite eggs.
2. As 1, plus significant removal of bacteria.
3. Not more than 100 coliform organisms per 100 ml in 80% of samples.
4. No chemicals that lead to undesirable residues in crops or fish.
5. No chemicals that lead to irrigation of mucous membranes and skin.

In order to meet the given health criteria, processes marked X X X will be essential. In addition, one or more processes marked X X will also be essential, and further processes marked X may sometimes be required.

The criteria recommended under recreation by WHO are equally applicable to irrigators who are likely to have physical contact with the effluent during irrigation.

Effluent irrigation may also lead to microbial contamination of air, soils and plants in the vicinity of the irrigation site. The extent of such contamination depends upon the degree of treatment provided, the prevailing climatic conditions, nature of the crop being irrigated and the design of the irrigation system. Where the terrain and the crop type are suitable, effluents may be applied through 'ridge and furrow' systems. These contaminate neither the air nor the upper parts of plants. Subsurface tile or trickle irrigation systems create the fewest hazards of any kind. However, the expense of utilizing such systems on a large scale severely limits their feasibility. An additional problem is the clogging of dripper nozzles and subsurface pipelines due to suspended sediments and microbial growth. Sprinklers create the greatest potential for microbial contamination of the vegetation and air.

When considering the use of effluents for irrigation, their microbial and biochemical properties will have to be evaluated. These values should then be

compared with the public health standards, taking into consideration the crop, soil and irrigation system and consumption of the produce, and only when the effluent meets these standards should it be evaluated in terms of chemical criteria such as dissolved salts, relative sodium content and specific toxic ions.

In quantitative terms, the volume of wastewater available for re-use by irrigated agriculture is negligible when compared with the overall volume of water used for irrigation. However, the potential impacts associated with water quality and agricultural re-use of wastewater are so important, economically, environmentally and socially, that the need for sound planning far exceeds the relatively small quantities and areas involved.

The following list of references contains research as well as practical information on various aspects of the re-use of effluents for crop production: Eckenfelder (1980); Loehr (1977); National Research Council of Canada (1974); Sopper and Kardos (1973); and Wilson and Beckett (1968).

SOIL AMENDMENTS AND IRRIGATION WATER QUALITY

The purpose of some soil amendments is to counter the effect of sodium, by increasing the soluble calcium content or by increasing the salinity of the irrigation water. Gypsum and other calcium supplying materials - Gypsum is the most commonly used soil amendment. Since water infiltration problems, caused by sodium, affect mainly the upper few centimeters of soil, repeated small applications of gypsum, incorporated at lower rates into a shallow depth, are preferred over a single large application.

If the salinity of the irrigation water is low (EC<0.5 ds/m), gypsum can be added to the irrigation water at rates of 1-4 meq/l of dissolved calcium.

Other amendments - when lime ($CaCO3$) is present in soil, some acids or acid-forming amendments can be used. These amendments cause calcium to be released to soil solution. Examples for such amendments are elemental sulfur, sulfuric acid and ferric sulfate.

Organic residues - these amendments improve soil structure and water infiltration, by keeping the soil porous.

Total dissolved salts

The total salt levelin irrigation water, or total dissolved salts (TDS), is usually assessed by measuring the electrical conductivity (EC) of the water. The common units of electrical conductivity are millisiemens per centimetre (mS/cm = dS/m) and μS/cm (1 mS/cm = 1000 μS/cm). The principle is that salty water conducts electricity and pure water does not.

Another unit sometimes used is milligrams of salt per litre (mg/L = ppm). This can be converted to EC by using a factor of 640: 1 mS/cm = 640 mg/L of TDS. (This conversion factor is approximate and depends on the actual salts present.)

Once you have a value for EC (from either your preferred laboratory or by measuring it yourself), you can use Table 7 to determine the suitability of the water for irrigation. (For more information, see How salinity is measured.)

TABLE 7. INTERPRETATION OF EC VALUES

EC	Class Interpretation
0 - 0.28	1 Low-salinity water suitable for use on most crops except on soils with very low permeability
0.28 - 0.80	2 Medium-salinity water posing few restrictions if reasonably drained soils allow some leaching. Under high evaporation and poor drainage, some precautions are needed
0.80 - 2.3	3 High-salinity water. Cannot be used on soils with poor drainage. Even with good drainage, leaching is essential, and only salt-tolerant crops can be grown
2.3 - 5.5	4 Very-high salinity water. Can be used only on well-drained soils with leaching, constant monitoring and salt-tolerant crops
>5.5	5 Extremely high-salinity water. Almost totally unusable even with strict precautions. Occasional emergency use in mild weather only

The total salinity of in the water is the most important determinant of use. Other important factors are the sodium content and the alkalinity.

Sodium

Sodium is not a plant nutrient and can be harmful to soils, causing clay dispersion and soil salinisation. It can also be harmful to plants when present in excessive amounts, causing tissue damage. EC or TDS tells us nothing about which salts are present. For example, a salinity of 1.0 dS/m due largely to calcium is less harmful than the same salinity due to sodium. The quantity of sodium is expressed as the sodium adsorption ratio, or SAR. This is the ratio of sodium to calcium plus magnesium: Na/?(Ca + Mg). SAR is described in Table 8.

TABLE 8. INTERPRETATION OF SODIUM ADSORPTION RATIO

SAR	Interpretation

<5	Few problems except under continued use with inadequate replacement of calcium as lime or gypsum
5 - 15	Increasing problems. Monitor soil salinity and exchangeable sodium levels regularly, and top-dress with calcium supplements or add soluble calcium to the water
>15	Almost unusable unless the water is modified by the addition of calcium

Alkalinity

Alkalinity is a further problem usually associated with sodium. Alkaline water can raise the pH of the soil to unacceptable levels, causing iron deficiency. This is made worse by poor drainage and salt accumulation. Soil must be regularly acidified (usually with agricultural sulphur or iron sulphate), or the water must be treated with acids. The usual unit of alkalinity is mg of lime-equivalent per litre.

The ideal water

The ideal water quality for irrigation has the following specifications:

- TDS < 175 mg/L
- EC < 0.28 dS/m
- SAR < 5
- Total alkalinity < 150 mg/L (depends on soil buffer capacity)

From the actual measurements we can make a good estimate of how the water will affect the crop and what management practices will be needed to control salinity build-up.

Controlling salinity build-up

Soil salinity control relates to controlling the problem of soil salinity and reclaiming salinized agricultural land. The aim of soil salinity control is to prevent soil degradation by salination and reclaim already salty (saline) soils. Soil reclamation is also called soil improvement, rehabilitation, remediation, recuperation, or amelioration. The primary man-made cause of salinization is irrigation. River water or groundwater used in irrigation contains salts, which remain behind in the soil after the water has evaporated.

The primary method of controlling soil salinity is to permit 10-20% of the irrigation water to leach the soil,that will be drained and discharged through an appropriate drainage system. The salt concentration of the drainage water is normally 5 to 10 times higher than that of the irrigation water, thus salt export matches salt import and it will not accumulate.

Although it is possible to use water with a wide variety of characteristics for irrigation, as water quality declines, management procedures become more demanding, and the requirements of the soils and plants become more restrictive. Failure to match the soil with the water and the management type can result in damage to soils and plants.

Water quality used in tank mixes for herbicides and pesticides

The calcium (Ca) content from various water sources will determine whether or not water is classified as hard, intermediate or soft. Hard water will have the highest level of calcium. The pH or acidity of the water has an influence on the chemicals mixed therein as well as the buffering capacity of the water being used. In simple terms the buffering capacity is the ability of the water to absorb increasing amounts of chemicals before there is a drastic change in pH which in turn is a measure of the number of available hydrogen ions in the mix. Neutral pH is measured at 7 and acid solutions are less than 7 with alkaline solutions being greater than 7.

BIBLIOGRAPHY

- Andersson, K., Rosemarin, A., Lamizana, B., Kvarnström, E., McConville, J., Seidu, R., Dickin, S. and Trimmer, C. (2016). Sanitation, Wastewater Management and Sustainability: from Waste Disposal to Resource Recovery. Nairobi and Stockholm: United Nations Environment Programme and Stockholm Environment Institute. ISBN 978-92-807-3488-1
- BN Puri, Irrigation and Agricultural Economy in Ancient India, Annals of the Bhandarkar Oriental Research Institute, Vol. 48/49 (1968), pp. 383-390
- Chartres, C. and Varma, S. (2010) Out of water. From Abundance to Scarcity and How to Solve the World's Water Problems FT Press (USA).
- Cohen, Yoram (2021). Advances in Water Desalination Technologies. Materials and Energy. Vol. 17. WORLD SCIENTIFIC. doi:10.1142/12009. ISBN 978-981-12-2697-7. ISSN 2335-6596. S2CID 224974880.
- David Gilmour (2007). The Ruling Caste: Imperial Lives in the Victorian Raj. Farrar, Straus and Giroux. p. 9.
- Desalination (definition), The American Heritage Science Dictionary, via dictionary.com. Retrieved August 19, 2007.
- Ebrahimi, Atieh; Najafpour, Ghasem D; Yousefi Kebria, Daryoush (2019). Performance of microbial desalination cell for salt removal and energy generation using different catholyte solutions. Desalination. 432: 1. doi:10.1016/j.desal.2018.01.002.
- FAO Water Unit | Water News: water scarcity. Fao.org. Retrieved 2009-03-12.
- Filley, S. How much does a cow need ? (PDF). Archived from the original (PDF) on 12 May 2012. Retrieved 17 March 2012
- Fischetti, Mark (September 2007). Fresh from the Sea. Scientific American. 297 (3): 118–119. Bibcode:2007SciAm.297c.118F. doi:10.1038/scientificamerican0907-118. PMID 17784633.

- for the historiography, see Rohan D'Souza, Water in British India: the making of a 'colonial hydrology'. History Compass (2006) 4#4 pp: 621-628. online
- Gleeson, Tom; Wada, Yoshihide; Bierkens, Marc F. P.; van Beek, Ludovicus P. H. (9 August 2012). Water balance of global aquifers revealed by groundwater footprint. Nature. 488 (7410): 197–200. Bibcode:2012Natur.488..197G. doi:10.1038/nature11295. PMID 22874965. S2CID 4393813.
- Gopal, Lallanji (2008). History of agriculture in India, up to c. 1200 A.D. p. 762. ISBN 978-81-8069-521-6.
- Gordon L., D. M. (2003). Land cover change and water vapour flows: learning from Australia. Philosophical Transactions of the Royal Society B: Biological Sciences. 358 (1440): 1973–1984. doi:10.1098/rstb.2003.1381. JSTOR 3558315. PMC 1693281. PMID 14728792.
- Haie, Naiṃ (2020). Transparent Water Management Theory: Sefficiency in Sequity (PDF). Springer.
- Hoiberg, Dale (2000). Students' Britannica India. Encyclopædia Britannica. p. 260. ISBN 978-0-85229-760-5.
- Husain Siddiqui, Iqtidar, and Husain Siddiqui, Iqtidar. Water Works and Irrigation System in India During Pre-Mughal Times. Journal of the Economic and Social History of the Orient 29 (1986): 52–78
- Ian Stone, Canal Irrigation in British India: Perspectives on Technological Change in a Peasant Economy (2002) pp 278-80
- India - History of Irrigation FAO - United Nations (2014)
- Making the Deserts Bloom: Harnessing nature to deliver us from drought, Distillations Podcast and transcript, Episode 239. Science History Institute. March 19, 2019. Retrieved 27 August 2019.
- Molden, D. (Ed.) (2007) Water for food, Water for life: A Comprehensive Assessment of Water Management in Agriculture. Earthscan/IWMI.
- Neil Charlesworth, British Rule and the Indian Economy, 1800-1914 (1981) pp 23-37

INDEX

T

V

W

Y